St. Louis Community College

Forest Park
Florissant Valley
Meramec

Instructional Resources
St. Louis, Missouri

GAYLORD

Every Living Thing

Daily Use of Animals in Ancient Israel

Every Living Thing

Daily Use of Animals in Ancient Israel

Oded Borowski

ALTAMIRA
PRESS

A Division of Sage Publications, Inc.
Walnut Creek • London • New Delhi

For information contact:

AltaMira Press
A Division of Sage Publications, Inc.
1630 North Main Street, Suite 367
Walnut Creek, California 94596 U.S.A.
explore@altamira.sagepub.com

Sage Publications Ltd.
6 Bonhill Street
London EC2A 4PU United Kingdom

Sage Publications India Pvt. Ltd.
M-32 Market
Greater Kailash 1
New Delhi 110 048 India

PRINTED IN THE UNITED STATES OF AMERICA

Library of Congress Cataloging-in-Publication Data

Borowski, Oded.
    Every living thing: daily use of animals in ancient Israel / Oded
Borowski.
            p. cm.
    Includes bibliographical references and index.
    ISBN 0-7619-8918-8 (cloth: acid-free paper).
    ISBN 0-7619-8919-6 (pbk.: acid-free paper)
    1. Domestic animals—Palestine—History. 2. Wildlife utilization—
Palestine—History. 3. Animal remains (Archaeology)—Palestine.
4. Human-animal relationships—Palestine. 5. Animals in the Bible.
I. Title.
SF55.P26B67 1997
636'.00933—dc21   97-21228
            CIP

98   99   00   01   02   03   10 9 8 7 6 5 4 3 2 1

Production and Editorial Services: David Featherstone
Editorial Management: Erik Hanson
Cover Design: Joanna Ebenstein

Contents

List of Illustrations

About the Author

ODED BOROWSKI is associate professor of Hebrew language and biblical archaeology in, and the former department chair of, the Department of Middle Eastern Studies at Emory University in Atlanta. Borowski received college degrees from the Midrasha/College of Jewish Studies and Wayne State University in Detroit before obtaining his Ph.D. in Near Eastern studies from the University of Michigan in 1979. For over twenty-five years, he has participated in the excavations of key archaeological sites in Israel, including Tell Gezer, Tell Dan, Ashkelon, Beth Shemesh, and Tell Halif, for which he currently serves as codirector of the project.

Professor Borowski is author of *Agriculture in Iron Age Israel* (Eisenbrauns, 1987). His research and writing have also appeared in *Biblical Archaeological Review, Biblical Archaeologist, Bulletin of the American Schools of Oriental Research, ASOR Newsletter, Israel Exploration Journal, Qadmoniot, Hebrew Computational Linguistics*, and *Am Va-sepher*, as well as in several books and encyclopedias.

Borowski has taught at Emory University since 1977. In addition, he has served at various times as Annual Professor, Dorot Professor, and acting director at the W. F. Albright Institute of Archaeological Research in Jerusalem. He has received grants for his research from the National Endowment of the Humanities, Georgia Endowment for the Humanities, Ford Foundation, Zion Research Fund, Memorial Foundation for Jewish Culture, Dorot Foundation, Emory University Research Committee, United States Information Agency, and others. He is actively involved in several organizations and publications that bring information about biblical archaeological information to the general public.

Preface and Acknowledgments

God blessed them and said to them, "Be fruitful and increase,
fill the earth and subdue it, have dominion over the fish in the sea,
the birds in the air, and every living thing that moves on the earth."
Genesis 1:24

When discussing ancient agriculture, the underlying tendency among scholars has been to deal with plants such as field crops, fruit trees, and the like. A good example is Kent V. Flannery's widely quoted article "The Origins of Agriculture" (1973), in which, under the term *agriculture*, the author discusses the beginnings of plant cultivation in Southwest and Southeast Asia, Mesoamerica, and the Andes without making any reference to animal domestication and husbandry and the possible relationship between these two aspects of agriculture. Another example is Mark Nathan Cohen's *The Food Crisis in Prehistory: Overpopulation and the Origins of Agriculture* (1977), in which the discussion concentrates on plant domestication and the conditions that led to it, while the domestication of animals is mentioned only in passing.

In my previous work, *Agriculture in Iron Age Israel* (1987), I took the same approach, and I was criticized for it by at least one reviewer of the book (Hadley 1988). In my defense, I would say that when I was writing that book, I was well aware of this dichotomy; however, the problem is one of nomenclature, and had I named the book in Hebrew, I would have replaced *agriculture* with the well-known term *ʿăbodat ʾădāmâh*, ('tilling the land, land cultivation'), which would have been much more precise. Using *agriculture* in the title of the book created, among English speaking readers, the expectation of the inclusion of animals as well as plants. I should have clarified this issue in the introduction and explained my intention of coming back at a later date with another volume dealing with animals in biblical times. While the present volume intends to fill this gap left by the previous one, it is not re-

13

stricted just to the place animals occupy in agriculture, but contains material about the use of animals in all aspects of daily life in biblical Israel. It deals not only with domestic, but also with wild animals, all of which were utilized by the Israelites. Furthermore, it looks at the use of animals in secular as well as cultic contexts. The presentation of the animals, which is done according to categories of usage, includes a brief history of their domestication and a discussion of their relationship with mankind. These are supported with biblical and extra-biblical references and are illustrated by references to ancient artistic representations depicting the animals and the different ways in which they were utilized. It was impossible to reproduce all the artwork, however, and the reader will need to examine most of the illustrations in the cited sources.

One purpose of this book is to clarify some misconceptions concerning the roles of the animals and of those who kept them in the Ancient Near East. The relationship of the farmer and the herder is most often portrayed as one of adversaries fighting each other, and it seems that many authors believe that the desert had an unrestricted ability to support a countless number of human beings. Unfortunately, this approach has been adopted by many who deal with the Near East but who are not historians. One example is the following, taken from *Palestine—Land of Promise* by Walter Clay Lowdermilk.

> Aside from the constant wars between external powers competing for control of this fertile and important crossroads, Palestine has suffered time and again from infiltration, invasion and raids by desert nomads. This age-old struggle between the shepherd and the farmer has made the Cain and Abel story of the Bible terribly real in the Near East for thousands of years. The economies of grazing and farming have always been antagonistic, and the desert has always produced more people than it can feed. In times of drought, hardy and desperate marauders have often swept out of the desert and ravaged cultivated regions. Sometimes they even supplanted the former population of the cultivated valleys and became farmers themselves, only to be driven out at some later time by a new horde invading from the desert. (1944:67.)

Needless to say, this is not the case, and although sometimes there was a certain tension between farmer and herder, there was much more cooperation between the two than strife.

To avoid misunderstanding, the historical-geographical terms used throughout the text need some explanation. *Syria-Palestine* or just *Palestine* is a geographical term referring to the area contained within the modern states of Syria, Lebanon, Israel, and Jordan. The name *Canaan* is applied to this area in the period before the appearance of the Israelites. *Eretz Yisrael* is a term used to define most of the region mentioned above during the Israelite period, from the time of the tribal settlement (ca. 1200 B.C.E.) through the fall of the Solomonic Temple (587/6 B.C.E.). As for the period covered by this book, although I use the broad term *biblical times,* I concentrate mainly on the Israelite period (Iron Age I and Iron Age II), taking into consideration earlier periods such as the so-called Patriarchal period, which presently cannot be dated.

Many institutions and individuals have helped me in carrying out this project, starting with the research and continuing through writing the results. Thanks are due to Emory University for granting me a research leave, the Memorial Foundation for Jewish Culture for a post-doctoral research grant, the W. F. Albright Institute for Archaeological Research (Jerusalem) for awarding me the Annual Professorship 1995–96, and the Joe Alon Center for Regional and Folklore Studies and Kibbutz Lahav for their general support. The completion of the project could not have been possible without the help of many friends, especially Mitch Allen, Brian Hesse, Liora Kolska-Horwitz, Paula Wapnish, Ed Wright, and others who helped with encouragement, reading parts of the manuscript, providing information, making suggestions, and answering questions. Special thanks are due to David Featherstone for his help with bringing the manuscript to its final shape. However, any mistakes are mine alone.

I would like to take this opportunity and thank my wife, Marcia, and my children, Jonathan and Orly, for their continuous support of my work and for enduring long periods of my absence.

Introduction

This book is dedicated to the study of the role animals played in the daily life of the people in the Ancient Near East, especially those living in biblical times in Eretz Yisrael. Animals—including birds, fish, and insects, as well as mammals—will be introduced according to their function (e.g., food, or burden) and their evolutionary relationship (species). Each presentation includes a description of the place the animal occupied in the economy, its place in human history as a result of domestication, and its by-products, such as milk and wool. It should be remembered that animals changed roles over time, since they might have been domesticated for one purpose but were later exploited for another. Also, the fact that animal species ate different kinds of plants enabled people to better exploit the vegetation in their environment.

The general question of the domestication of animals is briefly dealt with first. Since the physical evidence is varied, the nature of zooarchaeology—the study of bone remains—is examined and its function is explained. In exploring aspects of the relationship between the Israelites and animals, biblical references are presented not as proof of biblical historical accuracy but to illustrate the point under discussion. My assumption is that biblical traditions, no matter which period they describe, could not have survived if they were not tied in to real life. Many of the biblical references are examined in light of other textual evidence from Egypt or Mesopotamia since ancient records kept by the pharaohs of Egypt and the kings of the various Mesopotamian powers form one resource for the livestock inventory of Canaan (later Eretz Yisrael). The Egyptians, it seems, were mostly interested in taking home horses and chariots, while other animals were plundered mostly to feed the army. The Assyrians, on the other hand, were willing to take home any animal that could be utilized in one way or another. Further examples of the livestock inventory are found in artistic representations extant in the neighboring ancient cultures. Because Eretz Yisrael is relatively lacking in artistic representations, most of the illustrations of pertinent subjects come from other cultures with richer artistic legacies. Israelite daily life, like its material culture, was significantly impacted by outside influences, thus it is possible to illustrate it by using examples from the cultures in which they originated. Moreover, many customs and modes of life related to the daily use of animals in biblical times are illuminated by observing similar situations in more contemporary preindustrial societies and ethnographic evidence that can be gleaned from them.

∼

Since most of the animals dealt with in this work were exploited as food, let me start with a few words concerning the dietary importance of four-legged animals, fowl, fish, and insects. We know something about which animals the Israelites were ritually permitted to eat from examining the cultic dietary instructions contained in Leviticus 11 and Deuteronomy 14. However, one must remember that there is no way to know when these rules were formulated, or how closely they were followed and observed. In general, wild and domesticated herbivorous mammals could be eaten with certain restrictions. To be consumed, an animal had to have split hoofs and chew its cud. Animals that fulfilled only one of these requirements, and thus could not be eaten, are mentioned by name. Eating camel, hare, and hyrax was not allowed because they do not have split hoofs; pig could not be eaten because it does not chew its cud. To make the identification easier, Deuteronomy 14:4–5 contains a list of domesticated and wild animals that could be eaten. The list contains ox, sheep, goat, and a roster of seven wild animals, including deer or ram, gazelle, roebuck, and others. From the number of times they are mentioned, it appears that the most common domestic animals eaten were small cattle—that is, sheep (*kebeś*, or *keśeb*) and goats (*ʿēz*).[1] Their young ones (*śeh*, and *gĕdî*, respectively), were considered the choicest. Large cattle, (*bâqâr*) were also eaten, but not as often. This probably has to do with factors such as the size of the animal, the number of eaters, the function the animal had in the household, the number of animals available, the occasion for the meal, the problems of processing the meat and preserving it, and more.

Fish (*dāgâh*) were consumed, but little is known from the Bible about those that were eaten except that they had to have fins and scales. Nevertheless, when fish and molluscs can be identified in archaeological contexts, we can learn much about trade relations by determining their place of origin, which can be well defined. Birds are referred to in the Bible by the collective term *ʿôp*, fowl; however, three are specifically mentioned as permitted for consumption—quail (*śĕlāw*), dove (*yônâh*), and turtledove (*tôr*). Birds of prey and scavengers are prohibited from being consumed. The list identifying the unclean birds is long and appears in Leviticus 11 and Deuteronomy 14. Bird eggs were also a dietary component, but the Bible mentions them only in alluding to the collection of eggs of wild birds. Insects of the grasshopper family could be eaten, but all other insects were prohibited. Also prohibited were lizards, rats, mice, and turtles.[2]

Those who were not personally engaged in raising animals, hunting, or fishing, could obtain meat of different kinds in the markets of cities and towns. One such market, dated to the period just before the destruction of the city in 604 B.C.E. by the Babylonians, was uncovered in Ashkelon (Hesse and Wapnish 1996). A separate area was set aside in the market for carcass processing. Certain inedible parts were removed there, and the rest of the carcass was taken to the butcher's shop for sale. Bones that could be used for the manufacturing of tools or other objects were also removed in the preliminary process. The zooarchaeological evidence suggests that shops specialized in particular types of meat. Cows were sold in shops separate from those where sheep and goats were sold. Birds and fish were also sold separately. A small number of game animals were also sold in this market, suggesting that a segment of the population was engaged in hunting and people in the urban community were willing to buy and consume venison.

How was meat eaten in biblical times? Again we can learn much about meat consumption from the prohibitions. Meat was not eaten raw, and there was a strict prohibition against blood consumption. From the incident described in 1 Samuel 14:32–34, where Saul's men ate captured animals without draining the animals' blood, we learn that the prohibition was well-known

but not always observed. The question remains then, how strictly it was observed. Don and Patricia Brothwell note that "blood was not only utilized as
food when beasts were slaughtered, but…in all probability the drawing of
blood from live animals was practiced by some earlier cultures" (Brothwell
and Brothwell 1969:47–8). Against this custom, the Bible prohibits eating
meat with the blood: "But you must never eat flesh with its life still in it, that
is the blood" (Gen 9:4). Meat of a dead (*něbēlâh*) or a devoured (*těrēpâh*)
animal was not permitted to be consumed by humans, yet it could be fed to
dogs. Meat not fit for consumption was known as *piggûl*.

There were several ways of preparing meat in ancient Israel.[3] One way
was to boil it in water (*měbuššāl běmayim*) in a large pot; a by-product of
boiling meat was broth (*mārāq*). However, there was a strict prohibition against
cooking a kid (young goat) in its mother's milk (Milgrom 1983:112–3; Ratner
and Zuckerman 1986; Milgrom 1991:737–42), although the source of this
prohibition is debatable. Meat was roasted on a fire (*șělî ʾēš*), but we do not
know how common it was, since this method of food preparation is mentioned only a very few times, twice in connection with the Passover sacrifice.
Nevertheless, its mention in Isaiah 44:19 suggests that roasted meat might
have been part of the daily diet.

In addition to cooking in pots on an open fire, food was also prepared
in an oven (*tannûr*) or on a hearth (*kirayyim*), in a griddle (*maḥăbat*) or pan
(*marḥešet*). Food was served in dishes (*șallaḥat*) or bowls (*șělôḥît*), but no
eating utensils are known from the biblical period. Eating habits included
picking the food with the fingers and wiping the food from the dish with a
piece of bread in a manner similar to what is still done today in the Near East.
The latter habit serves as metaphor for the prophet: "I will wipe Jerusalem as
one wipes a dish, wiping it and turning it upside down" (2 Kgs 21:13).

There are ample hints in the Bible concerning which parts of the animal were considered the choice parts. In 1 Samuel 9:24 we find that the thigh,
most likely the front right thigh (*šôq hayyamîn*), was reserved for honored
guests. The priests also received this thigh, as well as the breast (*ḥazeh*).
These parts were pulled out of the cooking vessel with a fork (*mazlēg*), probably a large, three-pronged one.

Some animals were specially fattened for consumption. These included
calves (*ʿēgel marbēq*), sheep (*kâr*), ox (*měrîʾ*), and most likely "fattened ram"

(*ʾayyil millûʾîm*). Birds were also fattened, as is mentioned in the food list prepared for Solomon's table, which includes *barburîm ʾabûsîm*.

The menu of a typical meal is included in the story of Sinuhe (ca. 20th–19th century B.C.E.), the Egyptian nobleman who visited Canaan during the pre-Israelite period: "Bread was made for me as daily fare, wine as daily provision, cooked meat and roast fowl, beside the wild beasts of the desert, for they hunted for me and laid before me, beside the catch of my (own) hounds. Many…were made for me, and milk in every (kind of) cooking" (Pritchard 1969a:20). The menu of an Israelite meal for special occasions is presented several times in the Bible, one of which is in the story describing the visit of the divine messenger to Gideon, in which he "went in, and prepared a young goat and made an ephah of flour into unleavened bread. He put the meat in a basket, poured the broth into a pot . . . " and served the guest (Judg 6:19).

According to biblical prescriptions, the role of animals, mostly domestic, in the cult of Israel was central. Besides the fact that the priestly diet depended to a certain degree on parts resulting from the sacrifices, there were several occasions in which animals were either completely consumed by the sacrificers, as on Passover, or by the fire on the alter, as in guilt-offering. Some information concerning this role is provided by zooarchaeological studies. This information can help determine the cultic use of wild and domestic animals, as well as that of herbivores and carnivores.

Notes
1. The term *small cattle* is a designation used to refer to sheep and goats, while *large cattle* is used for cows, oxen, and similar animals. While these terms are widely used, I have been unable to determine their origin.
2. On the dietary laws, see Milgrom 1983:104–18.
3. Most of the prepared meat came from domestic animals such as sheep and goats, although even in urban centers some of the meat eaten by the inhabitants came from wild animals such as gazelle and fallow deer (Kolska-Horwitz and Tchernov 1989). Cut marks discernible on some bones, demonstrate butchering techniques and illustrate their consumption. The appearance of gazelle and fallow deer at an earlier period (Iron Age I) in a provincial settlement such as Beer-Sheba (Hellwing 1984:110) should not be surprising.

Chapter 1

The Story of Animals

DOMESTICATION OF ANIMALS

Domestication is a totally different approach to animal management than hunting-gathering, since it requires a decisive action for people to manipulate nature rather than merely to take advantage of it. Before considering how animals were used on a daily basis during the Iron Age, however, a few questions need to be raised. The first is why and when humans became engaged in animal domestication. Following this, we can ask how man went about domesticating animals and in what order it was done. How did domestication manifest itself, and what were the effects on man, on the animals, and on nature as a whole? Several scholars have been looking into all or part of these questions, and the results of their inquiries are used in this publication (Brothwell and Brothwell 1969; Bulliet 1975; Cohen 1977; Clutton-Brock 1981; Davis 1987; Uerpmann 1987; Bökönyi 1989; Anthony 1991; Grigson 1995).

Sometime at the end of the Paleolithic period,[1] man became more proficient with the tools he was manufacturing. This resulted in changes to the diet; with more flesh included, and the population may have increased to over three million. Further technological advances during the Mesolithic period may have caused the increase to continue (Brothwell and Brothwell 1969:15). This population growth was the cause for, rather than the result of, much of the great change in subsistence patterns. The present evidence suggests that agriculture, and later the domestication of livestock, was adopted in a few restricted areas where the carrying capacity of the land could no longer support a stable population of hunter-gatherers (Clutton-Brock 1981:189). Both the domestication of animals and the development of agriculture were actually responses to the increasing numbers of human populations; since this growth took place in more than one location, domestication therefore occurred independently in several areas rather than being dispersed out of one center (Cohen 1977:24).

Recent studies have shows that the diet of hunting and gathering populations was much better than that of agriculturists. Hunting and gathering activities provide a wider variety of foods than agriculture can, and the food resources for hunters and gatherers may be more reliable than agricultural resources. Furthermore, agricultural production is much more labor intensive than hunting and gathering and is more prone to failure because of climatic conditions (Cohen 1977:27). There is nothing to suggest that people would shift to agriculture to save labor or to gain leisure time since, in the transition to agriculture, labor costs increased rather than decreased (Cohen 1977:34). Hence, it appears that even though sedentism and its reliance on domesticated plants and animals is not necessarily the most desirable situation, it became the norm because of need rather than by design.[2]

The advantage of sedentism and domestication over hunting and gathering is that it produces a greater number of total calories for each unit of space in a given unit of time. Domestication techniques were developed that capitalized on species that might, although palatable, had the capacity to provide large quantities of storable calories or storable protein in close proximity to human settlements (Cohen 1977:279–280).

From all the above, it appears that there is a revolving cause and effect relationship between domestication of plants and animals and of population growth. Population growth forced mankind into better control of food resources through domestication and agriculture, and this in turn encouraged further increase in population and life expectancy. Taking control over food resources became a growing necessity, and by taming and domesticating animals man created a "walking larder" (Clutton-Brock 1981:20, 190).

Scholars agree that domestication is a process that takes many animal generations and relies on both animal availability and human needs. It probably takes centuries to transform a wild animal into a domestic one, and the pace varies from one species to another. Early in the process, when the animal is still wild, products such as meat, milk, fiber, leather, and dung can be utilized. As the process continues, it becomes possible for the animal to be used for riding, bearing burden, or draught (Bulliet 1975:38).

Domestication involves the physical removal of the animal from its wild relatives—and at least occasional control of its breeding—for the benefit or profit of its controller (Anthony 1991:251–2). Bökönyi stresses the fact that animals to be domesticated are of a species with particular behavioral characteristics, and their maintenance and breeding conditions are totally controlled "for mutual benefits" of both mankind and the animals (Bökönyi 1989:22).

Ducos defines domestication with reference to human society and the place animals occupy in it. "Domestication can be said to exist when living animals are integrated as objects into the socio-economic organization of the human group, in the sense that, while living, those animals are objects for ownership, inheritance, exchange, trade, etc., as are the other objects (or persons) with which human groups have something to do" (as quoted in Bökönyi 1989:23).

Clutton-Brock agrees that "a domestic animal is one that has been bred in captivity for purposes of economic profit to a human community that maintains complete mastery over its breeding, organization of territory, and food supply." However, she points out that "not all adult mammals can flourish or will breed under such drastic alterations to their natural way of life, although all young mammals can be tamed when nurtured under the right conditions" (1981:21).

The domestication process is gradual and dynamic, but not always irreversible. Escaped domestic animals can become feral and revert to some of their old modes of behavior, but the process of domestication itself tends to change the domesticated animal, first culturally and later physically. The process trains the animal to behave in a manner acceptable to the trainer and promotes certain desired physical qualities. Domestication leads to morphological changes such as decreased size, crowded teeth, and a hornless skull. About thirty animal generations are required after the beginning of domestication before measurable changes appear. In prehistoric sites, when a decrease of size in animal remains is detected, it can only indicate domestication. Similarly, crowded teeth and hornless skulls are never found in wild cattle, though they do occur rarely in populations of wild sheep; but when such remains are found among animal bones from an archaeological site, it is a true indication of domestication (Bökönyi 1989:25).

The morphological changes enumerated above[3] are caused by an imbalance and disruption in the rate of growth of different parts of the organism. How this happens is not well understood, but it is probably due to stress and to hormonal changes resulting from the animal's emotional and physical dependence on man (Clutton-Brock 1981:21–2).

Another outgrowth of domestication is the development of animal husbandry, which has two phases. The first is primitive animal-keeping done with unintentional, rather than conscious, breeding selection. The second, which involves animal breeding with conscious selection, mainly aims to increase productivity and can be recognized in the development of different breeds in a given population (Bökönyi 1989:26).

According to Clutton-Brock "a *breed* is a group of animals that has been selected by man to possess a uniform appearance that is inheritable and distinguishes it from other groups of animals within the same species" (1981:26). The process of selective breeding is probably as old as the Neolithic period and was practiced by the earliest farmers to produce animals that were "distinctive and submissive, as well as small, hardy, and easy to feed" (Clutton-Brock 1981:26). *Breed* should not be confused with *subspecies*, however. The difference between these two terms is that a breed is a product of artificial selection by man and geographical barriers need not play any part in its development, while a subspecies is always restricted to a given locality where it has evolved as a result of reproductive isolation (Clutton-Brock 1981:29).

What made domestication of animals possible? The conversion to a completely different method of food production could not have been achieved through isolated taming attempts, but required massive attempts at domestication. According to Bökönyi, large-scale domestication of animals and the Neolithic transition to domestic fauna occurred hand-in-hand with the beginning of cereal production. The latter provided the large amounts of rough fodder needed for the caprovines, or small cattle, the leading species of the first wave of animal husbandry at the beginning of the Neolithic period (Bökönyi 1989:23).

Domestication is only the first step in the symbiosis between man and animal. Through domestication, animals influence both man and society, although their influence is not as strong as that of man on them. Therefore, only animals with particular behavioral characteristics, like sheep and goats, can be domesticated. These species possess social behavior that enables their controlled breeding while other species, like the gazelle and onager, have some behavioral barriers that block their domestication (see also Clutton-Brock 1981:11; Bökönyi 1989:24).

In 1865, Francis Galton wrote an essay on the domestication of animals, saying in part, "I will briefly restate what appear to be the conditions under which wild animals may become domesticated: 1, they should be hardy; 2, they should have an inborn liking for man; 3, they should be comfort-loving; 4, they should be found useful to the savage; 5, they should breed freely; 6, they should be easy to tend" (in Clutton-Brock 1981:10).[4] Accordingly, only animals that can adapt to living and breeding within the human social system, and isolated from their wild relatives, can be domesticated. This means that, by nature, animals to be domesticated must conform broadly to the same kind of society, since they move all into a new ecological niche provided by the agriculturists (Clutton-Brock 1981:188, 190).

Scholars disagree on the ecological ramifications of domestication and on whether the types of animals domesticated in the Near East are the product of the environment or are its creators. According to Uerpmann, "during the Final Pleistocene, most of this region was barren as it is today" (1987:9), "and animals adapted to arid conditions are characteristic of its fauna" (1987:133). Clutton-Brock thinks otherwise.

> Five thousand years ago western Asia was a land of fertile plains, it was the hearth of civilization; today these plains are mostly desert, due to centuries of over-grazing by goats and sheep and to the relentless collection of every scrap of shrub and twig for firewood and the production of charcoal. Climatic change may also have hastened the production of deserts in the Middle East, as in the Sahara and elsewhere, but then it is not known to what extent the activities of man have, in turn, affected climate and rainfall. (1981:191–92.)

There is general agreement that domestication emerged in a rather late phase of mankind's history. It had Paleolithic and Neolithic antecedents in the form of the isolated taming and keeping of dogs[5] and pigs, two animal species that did not present any competition to man because they could live on the remnants and debris of human food and did not need large quantities of vegetable fodder (Bökönyi 1989:23). During the second half of the eighth millennium B.C.E., domestication of sheep and goats occurred at the vertex of the Fertile Crescent and south Anatolia, and quickly spread over the entire Proto-Neolithic region. A few centuries later, cattle and swine[6] joined the caprids as early domesticates. By that time the basic techniques of animal husbandry were already known to the potential domesticators through the earlier experience they gained by rearing sheep and goats (Uerpmann 1987:140). The primary reason for the domestication of all of these species was the same—the production of protein and fat (Uerpmann 1987:141).

It is quite possible that the dog predates sheep and goats in its association with man, dating from when tamed individual wolves participated with man in hunting animals. As the caprids became domestic, the dog then began to participate in herding.[7] When man started having other domestic animals, the wolf became an enemy, making it crucial for the man-adapted wolves to look different from their ancestors. This situation created selective pressures that favored the domestic features of the dog. Therefore, the early appearance of obvious dogs is closely related to the primary domestication of sheep and goats in the Near East (Uerpmann 1987:141).

The first animal to be domesticated for labor in the Near East seems to have been the donkey. There is an assumption that donkey domestication took place in Northeast Africa in the fourth millennium B.C.E., but the oldest hard evidence for the domestic donkey comes from Mesopotamia, while the earliest clear evidence for the use of the donkey in Egypt appears in a well-known depiction of a donkey rider dated to the Fifth Dynasty (ca. 2700 B.C.E.) (Uerpmann 1987:141 and fig. 62), which is somewhat later.

As for the horse, there is no clear evidence that the Near East played any role in its early domestication (Uerpmann 1987:142). This animal appeared on the domestic scene quite late (Clutton-Brock 1981:20).

The Bactrian camel and the dromedary seem to have been the latest major additions to domestic livestock. Summarizing available information concerning the these animals, Uerpmann maintains that there is not enough data for any clear statements about their early history. Nevertheless, the domestication of the dromedary seems to have taken place in Arabia, while the Bactrian camel may actually have been domesticated in Bactria and possibly in northeastern and eastern Iran, as well (Uerpmann 1987:142).

Economically, ungulate ("plant-eating") mammals have always been important to man (Uerpmann 1987:9). Animal remains indicate that, during the Chalcolithic and Bronze ages, the economy of the Near East was based on the breeding of four meat-producing ungulate species: sheep, goat, cattle, and pig. The same is true today, except where religious proscriptions or unsuitable environmental conditions prevent pig-keeping. That these animals were domestic is obvious, because by the Chalcolithic period all of them had undergone a marked decrease in size and there is no question that, apart from the infrequent hunted wild animal, all the remains of sheep, goats, cattle, and pigs represent domestic stock (Grigson 1995:247).

ZOOARCHAEOLOGY

The study of animal bones in archaeological context has been an expanding subdiscipline for the last forty years; nevertheless, the scope of this field is still being formulated (Hesse and Wapnish 1985:3; Davis 1987:23).[8] Even its name is still under discussion. Some refer to it as *archaeozoology*, and others as *zooarchaeology* (Legge 1978; Hesse and Wapnish 1985). Still others avoid these terms altogether, because they do not believe that they are doing zoology with archaeological materials or archaeology with zoological principles, but archaeology with one kind of residue—bone (Hesse and Wapnish 1985:3). While according to Davis, "Zooarchaeology is the field of study which helps recover the story of animals in prehistoric and historic times" (1987:3), Hesse and Wapnish see the emphasis "on human activity and how it may be inferred [from the study of bones]. Thus, the structure of analysis is cultural rather than biological" (1985:3).[9] For the sake of convenience, and without taking sides in this discussion, I will use the term *zooarchaeology*.

Being relatively young, the character of this field is still being molded. Its main problem, however, is not that but the fact that many archaeologists are still unfamiliar with the type of answers zooarchaeological analysis can

provide. The attitude toward the recovery of bones has been different than
that of other material-culture objects. Thus there are those who do not conduct a thorough retrieval operation, which, among other things, should
include sifting with small-mesh screens, flotation, and sampling of crucial
loci such as floors. Some of these concerns were expressed by Legge, who
said that "even a brief encounter with archaeological field work will show
the nonarchaeologist that animal bones (and, of course, other sorts of material) are recovered according to very different standards at various sites, with
the methods used often being determined by the period, and even the area,
within which the site is found" (1978:129).

Other problems confronting the zooarchaeologist stem from the nature
of the evidence. Being organic matter, most bones do not survive or survive
in a very poor condition (Fig. 1.1). Only under special circumstances are
bones preserved in a good state. Furthermore, most of the faunal material
found in archaeological excavations consists of mammal bones and teeth.
Bones of birds, reptiles, and fish, and shells of molluscs are generally found
in smaller quantities (Davis 1987:47), and their analysis requires experts and
special comparative collections, the number of which is much smaller than of
those dealing with mammals.

One of the problems in dealing with material from the earlier periods is
that of distinguishing "wild" animals from "domestic," a problem similar to
that of distinguishing between breeds. Very few fragments of horn or scraps
of hide are left that can be helpful in the matter of breeds, although at times
an occasional figurine found in close relationship to the skeletal remains may
be helpful by illustrating the animal under study. Artistic representations of
animals from the beginning of the second millennium B.C.E. provide evidence
that both the Mesopotamian and the Egyptian civilizations had developed
definitive breeds of dogs, cattle, and sheep (Clutton-Brock 1981:26).[10]

Nomenclature is another problem facing the experts. In early strata, it
is not unusual to find skeletal remains of animals, often in a very fragmentary
condition, that are difficult to describe since it cannot be determined whether
they represent animals that were wild, tamed, or in any other way exploited
by man. Formerly, these specimens were often identified as paleontological
"fossils" and were given new names according to the formal zoological code.

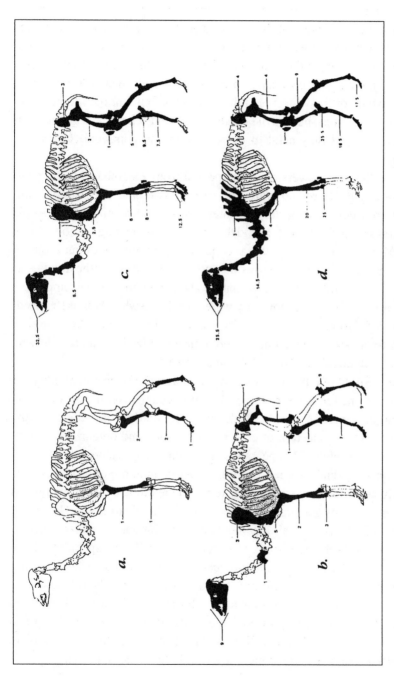

Figure 1.1. *Skeletal outlines illustrating an archaeological animal. The elements of camels (shown in black) encountered in refuse associated with four periods at Tell Jemmeh. (After Hesse and Wapnish 1985:fig. 87.) a. Bronze-Early Iron (1400–800 B.C.E.); b. Assyrian (800–600 B.C.E.); c. Persian-Babylonian (600–400 B.C.E.); d. Hellenistic (400–200 B.C.E.)*

Today, however, they are usually treated as evidence of initial domestication. The correct naming of these finds still presents difficulties, as does the naming of the later stages in the succession from the wild ancestor to the highly specialized animal that now lives in association with man (Clutton-Brock 1981:27). Many of the widespread discrepancies found in the literature are not real contradictions, but are the result of different schools and generations of researchers, and they need to be addressed and corrected (Uerpmann 1987:10).

What kind of answers can be expected from zooarchaeological studies? According to Hesse and Wapnish, there are two objectives guiding animal-bone archaeology, to portray the interaction between animals and people in a cultural setting and to understand the processes motivating the zoocultural system. Most zooarchaeologists are concerned primarily with questions of diet and environment. This focuses the interpretive effort on the animals as the unit of study and stresses information about how the animals appeared, where they could be found, how they were raised, and what their useful products included. However, "it is equally important in a fully anthropological approach to consider how animals were conceptualized by the people who interacted with them" (Hesse and Wapnish 1985:5).

For the field of biblical studies, it is important to be able to identify the animals mentioned in the Scriptures. Here we need to take into account folk identification, remembering that folk classification does not abide by any theory to direct the selection of criteria for defining and organizing categories. Elements of physical appearance are taken at face value; namely, they are descriptive and nothing more.[11] Folk classification might be responsible for mixing and confusing some of the terms in the biblical taxonomy related to herd animals (for general principles of folk classification see Hesse and Wapnish 1985:9). Historical documents and ethnographic reports concerning either the culture under study or groups temporally and spatially related to it, or at a similar level of technological development, may suggest or describe animals that were preferred, tabooed, used in rituals, or viewed as totems. These resources thus recognize culturally valid classes (Hesse and Wapnish 1985:6). For our purposes, the study of written and art-historical documents from the Ancient Near East, together with accounts from preindustrial societies in the region, are very helpful and important.

Skeletal elements themselves can provide information concerning the sex and age of an animal at the time of death. Age and sex data can help in determining whether the economic system was hunting and gathering or animal husbandry (Davis 1987:39, 44). For example, the early stages of animal domestication in the Pre-pottery Neolithic period can be determined in excavations where concentrations by sex and age of bones of specific types of slaughtered animals indicate that the early humans there had a certain amount of control over the age and sex of animals in their environment, something that cannot be achieved by hunters (Nissen 1988:24).

Age and sex data of domestic animals can also suggest the degree of involvement in animal husbandry. Usually, pastoralist flocks are divided into two groups, animals destined to be sold and animals required to maintain and increase the herd. Animal bone remains generally fall into three patterns, each reflecting a different kind of use. "A self-sufficient pastoral community should be associated with a mortality pattern closely comparable to the normal mortality of each herded species. A consuming community should be associated with an abundance of animals of marketable age. A group of producers involved in market economy should be represented by abnormal frequencies of sub-adults and females" (Hesse and Wapnish 1985:16). These patterns are based on the observation, that in many pastoral systems, the optimum slaughter age is toward the end of the juvenile period, when rapid growth stops and the meat gain is relatively small compared to fodder requirements. Most males are culled while young, and only a small number are kept, mostly for stud purposes; the females are kept longer, for reproduction and for such by-products as milk and wool (Davis 1987:39).

Another helpful category of data is related to the location of animal parts in various areas of the investigated site since the distribution of remains of certain cuts of meat can provide information concerning social classification and the economy. In the past, as in the present, certain parts of the animal were considered choice and were preferred by those who could afford them. The relative distribution of remains of these parts, and of the less-preferred ones, at particular loci may indicate the social and economic status of the segment of population residing there (Hesse and Wapnish 1985:13–4). In addition, the proper identification and quantification of the slaughtered animals can help determine how much meat was available during a particular period of habitation. An approximate meat weight of the four domestic species prevalent in the Near East in the Late Neolithic period has been proposed by Clark and Yi: Sheep, 80 kg; Goats, 80 kg; Cattle, 625 kg; Pig, 100 kg (in Grigson 1995:247–48). Accordingly, it has been suggested that in almost all such sites up to the Iron Age, cattle provided at least 50 percent of the meat (Grigson 1995:251).

Furthermore, zooarchaeological remains may provide information concerning the paleoenvironment. This is based on the assumption, which should be coordinated with paleobotanical and sedimentological studies (Hesse and Wapnish 1985:17), that the present-day dietary and climatic preferences of an animal were the same in earlier periods (Davis 1987:61). However, it should be remembered that a sample recovered from an excavation represents only the specific animals man selected from the environment, and not the nature of the total environment.

Another way of studying the paleoenvironment, especially microenvironments, is through parasites. At some stage of their life cycle, many types of parasitic worms in the gut of animals (including humans) develop resistant eggs or cysts that have a hard chitin-like covering (Davis 1987:73). Since some of these pass out with the feces, they can be studied as coprolites or through samples taken from the animal's abdomen cavity.

Environmental studies in general, and zooarchaeology in particular, need to consider biases introduced by the processes that affect the deposition and recovery of samples. Hesse and Wapnish (1985:20–31) identify them as follows:

- *Biotic* processes are those related to the environment which determine the species and their seasonality.
- *Thanatic* processes are responsible for the removal of members from a living population and depositing them in an archaeological context.
- *Perthotaxic* processes are those that move and destroy fragments of bones before their deposition and burial, including disarticulation, selective deposition, weathering, gnawing, and trampling.
- *Taphic* processes are related to mechanical and chemical actions.
- *Anataxic* processes are recycling processes through which bones are moved from their original burial place to a secondary one, as in the production of mud bricks.

- *Sullegic* processes involve biases introduced by the archaeologist because of all the above reasons and the selectivity of the digging process.
- *Trephic* processes introduce errors through curatorial factors such as sorting, recording, and reporting.

The results of the analyses can be more accurate by anticipating and identifying the sources and reasons for biases. Moreover, at times it would be better for the sample not to be studied because of the inability to correct the bias (Hesse and Wapnish 1985:31–32).

In the final analysis, the makup of the osteological data can help in determining the nature of the animal dependency of the community under study. The ups and downs of animal husbandry and the changes in its direction and emphases can be adduced from the zooarchaeological record (Fig. 1.2). Hence, zooarchaeology is a highly important field for biblical studies, whether for understanding a population's daily diet, economic system, cultural milieu, or cult, all of which influence historical processes and are reflected in the biblical text, whether in law codes, historical narratives, or prophetic utterings. To demonstrate this, I will use as an example a study conducted at Tel Dan showing the changes from the Late Bronze Age (LBA) to the early Iron Age (IA) that illustrate a possible response to a biblical question of the Israelite settlement, especially at this site (Wapnish 1993:430–5).

The LBA bone sample from Tel Dan contained 50% from cattle, 25% from sheep and goats; the rest belonged to fallow deer and a few pigs. The makeup of the sample illustrates a type of town economy supported by intensive agriculture and driven by the high use of cattle in plowing and other agricultural chores. The early IA sample was divided into two phases. The earlier phase contained 83% from sheep and goats, but only 17% from cattle. This sample is characteristic of a pastoral community relying only slightly on agriculture. However, in the sample from the later phase of the early IA, cattle increased to 49%, while sheep and goats decreased to 51%. This indicates a shift in the economy toward plow agriculture. Furthermore, it appears that as the number of sheep and goats decreased, the number of small cattle slaughtered in their second year increased, suggesting a reorientation toward meat production. Such a ratio could also reflect meat being importing from the outside, and analysis of slaughter offal indicates that, in the earlier phase,

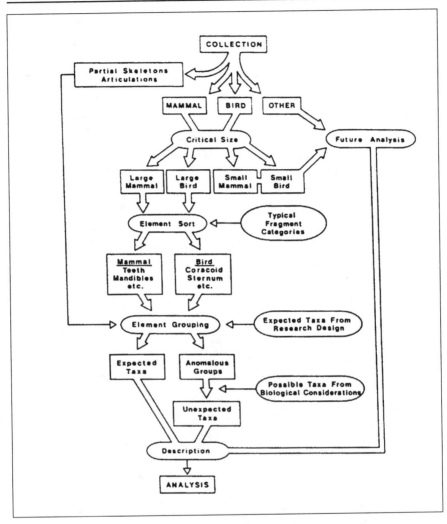

Figure 1.2. *Flow chart of the process of sorting in animal bone research. The double lines indicate the movement of material and the single lines the movement of information.* (After Hesse and Wapnish 1985:fig. 87.)

beef was acquired from the outside while in the later phase sheep and goats were imported and cattle were processed on the site. This analysis suggests that as time progressed, plow agriculture became more significant. Moreover, as more cattle were employed in agriculture, human labor was released to be used in other projects, such as public construction or herding at a distance from town (Wapnish 1993:430–5).

The picture emerging from this study shows the economic evolution of the Dan community as it changed from agricultural based in the LBA to pastoral in the early IA and then evolved back to agricultural. The pastoral community of the earliest phase of the IA could not have originated in an agrarian culture somewhere in the lowlands, but rather had a pastoral background. Since the time span covered by this study is considered by most scholars as that of the Israelite settlement, the observed process cannot fit the biblical model that suggests the Israelites came out of an agrarian or urban background as part of a peasant revolt. It seems, rather, that the settlers of the first IA phase had a very strong pastoral history.

Notes

1. The end of the Pleistocene was the period of incipient agriculture; it was also the time of the transition between the tamed wolf and the domesticated dog (Clutton-Brock 1981:188).

2. While many archaeologists believe that the people who first domesticated plants and animals, especially in the Near East, had already been leading a sedentary life for one or more millennia, it appears that mankind was forced to become technologically innovative as a result of a population increase and the threat of starvation (Davis 1987:76).

3. Clutton-Brock sees the sequence of general effects and signs of domestication as follows: I. body size; II. outward appearance and pelage characteristics; III. internal characteristics and dentition; IV. behavior; V. castration (Clutton-Brock, 1981:22–5).

4. For her interpretation, see pp. 15–6.

5. There are indications that in the ninth millennium B.C.E., Natufian people in the Near East already kept wolves (Uerpmann 1987:141).

6. Earlier, individual pigs were tamed, but for quite some time the pig was not a domestic animal. Pigs had almost certainly been domesticated by about 6000 B.C.E., and the domestication of cattle, which probably began in the sixth millennium, had resulted in a definite diminuation in size in the Levant by the fifth millennium (Grigson 1995:247).

7. This association is suggested by a dog mandible found in Palegawra cave in the same levels with the remains of domestic goats dated to the Early Neolithic (7th millennium B.C.E.) (Uerpmann 1987:141).

8. Because many archaeological excavations conducted in the first half of this century did not record and report information useful to this field, it is often impossible to assess the zoological and/or chronological reliability of what is included in the literature prior to 1950 (Uerpmann 1987:11).

9. Additionally, to fully appreciate the place of animals in human history, the study of paleoecological changes is necessary to account for changes in distribution and other shifts (Dayan, Simberloff, et al. 1991; Dayan 1996).

10. On the importance of using art-historical sources, see Hesse and Wapnish 1985:11–2.

11. An example of this is the lists of clean and unclean animals.

Chapter 2

Ruminants

Ruminants are wild or domestic herbivorous (plant-eating) mammals that ruminate; that is, they chew their cud. Unlike carnivores (meat eaters), these animals live in herds. Domestic ruminants include sheep and goats, which are collectively referred to as small cattle; another group is that of large cattle, which includes cows and other bovides. These animals are kept as a source of human livelihood through herding.

HERDING AS A WAY OF LIFE

Herding was prevalent in the Ancient Near East during the early stages of Israelite history. It was a way of life in the pre-Settlement (Patriarchal) and Settlement periods, and continued in the period of the Monarchy, during which Israelite urban life and long-distance trade developed. Although the extent of herding in the monarchical period cannot be estimated, the mention of the Sons of Rechab (Jer 35:1–11) is a good example for the persistence of a herding lifestyle during Iron Age II and as late as the destruction of the First Temple. As is apparent from their appearance in the Book of Jeremiah, in the beginning of the sixth century B.C.E. the Rechabites were still observing the dictates of the founding father of their clan to live in tents and not to plant or sow seeds. While herding is generally associated with desert nomads, it would be wrong to assume that it took place only in open spaces such as the Judean desert and the Negev. There is much zooarchaeological evidence showing that herding was also practiced in urban and semi-urban centers.

Types of Herding Systems

There are three production systems in present-day Near Eastern herding societies—sedentary, transhumant, and nomadic (Yalçin 1986:10–3). The type of production system in a certain society or locale is determined by "availability of forages, seasonality of vegetation and topographical and climatic conditions" (Yalçin 1986:10). An examination of the textual and archaeological evidence shows that all three of these production systems were practiced in ancient Israel.

In the sedentary system, herds are accompanied by shepherds and are kept at or close to the permanent settlement at all times. Grazing takes place during the day in the common (*migrāš*, Num 35:3; Jos 14:4)[1] or private grounds, which include cultivated fields. At night, the animals are kept inside the settlement. Flocks, which may be privately owned or village-owned, vary in size up to three hundred animals. In the winter, the animals are kept in shelters when climatic conditions required it (Yalçin 1986:11).

Herding under the sedentary production system was practiced by the native Canaanite population before the Israelite Settlement. This is reflected in the Conquest stories describing the destruction inflicted by the Israelites: "Then they devoted to destruction by the edge of the sword all in the city . . . oxen, sheep, and donkeys" (Jos 6:21). Israelite involvement in sedentary herding is suggested by many of the stories depicting village life. When Gideon, who lived in the village of Ophrah, hosted the divine messenger and offered him food, a young kid was readily available to be slaughtered (Judg 6:19).[2] The same was true when Manoah, who lived in the village of Zorah, hosted the divine messenger and offered him a meal that included a readily available kid (Judg 13:15).

The prevalence of the sedentary system is suggested by another story. When David of Bethlehem was young, he was shepherding his father's flock (1 Sam 16:19; 17:34). The flock apparently grazed somewhere near the village, which explains how Samuel's demand to bring David home from tending the flocks could have been fulfilled immediately (1 Sam 16:11). Furthermore, when Saul destroyed the town of Nob in retaliation for its helping David, he killed numerous animals sheltered in the town, "oxen, donkeys, and sheep, he put to the sword" (1 Sam 22:19). Maintaining herds inside the settlements continued even in the time of Hezekiah, when "[T]he people of Israel and Judah who lived in the cities of Judah also brought in the tithe of cattle and sheep. . . " (2 Chr 31:6). The countless references, of which only a few are offered here, demonstrate that the mode of life in the Israelite villages and towns included tending herds.

Archaeological finds of numerous sheep and goat bones at sites from the biblical periods add support to these textual descriptions. Additionally, architectural remains of structures that are interpreted as animal pens have been found in hill-country sites such as Giloh, near Jerusalem (Fig. 2.1), in the Negev Highlands, and at many other locations. Site plans in villages and towns containing houses adjacent to each other forming an outer ring, with a

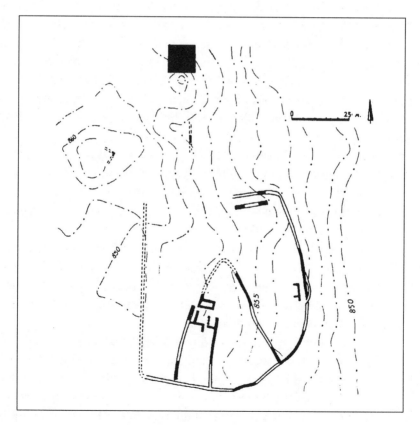

Figure 2.1. *Giloh, plan of the settlement.* (Mazar 1990:fig. 8.19.)

wall surrounding an open space with one entrance, strongly suggest that these were settlements where herding was practiced. The open space was most likely used for keeping herds at night and during foul weather or other bad conditions. Sites such as Early Iron Age Beersheba in the biblical Negev (Herzog 1984:figs. 34–5), 'Izbet Sartah in the central part of Israel (Finkelstein 1986:fig. 26),[3] and the so-called fortresses of the Negev were prime sites with herding as part of the mode of life.

Similar herding patterns also can be observed in Iron Age II sites. Zooarchaeological studies at Tell Halif, in the southern Shephelah, indicate that animals raised there were used for meat, milk, and wool. Cattle were used for labor, meat, and possibly milk. A comparison of the ratio of sheep to goats through the history of the site points out that it was 1 to 1 only in the Late Bronze Age and that in the Arab period it was 0.6 to 1. In all other periods, sheep were favored over goats, and in the Iron Age the ratio was 1.9 to 1.[4] The ratio of small cattle to large cattle was always in favor of small cattle. During the Iron Age, it was 5.5 to 1 (Zeder in Seger 1990:25–6). It appears that climatic conditions were not harsh, and the economy relied mostly on the by-products of sheep while utilizing cattle for labor, either in the fields or for transport.

The transhumant herding system demands that flocks, accompanied by shepherds, move from their home to other regions to take advantage of climatic conditions affecting temperature, humidity, and grazing conditions. When conditions improve at home, the herds return to graze or be fed there. These herds can be private or they can be communal, with several owners. In the latter case, the owners' contributions to the expenses depend on the number of animals they have in the flock. Transhumant flocks are usually larger than sedentary ones, ranging from 200 to 500 animals. They traditionally move from one region to another by walking (Yalçin 1986:11–2), although today flocks can be transported by trucks.

While the transhumant system might have been the most popular in ancient Israel, there is only one good example to illustrate it. This example is related to the early history of Israel, the Patriarchal period. While according to biblical descriptions Jacob's ancestors were nomadic herders, Jacob seems to have become acquainted with transhumance in Padan Aram, in northern Mesopotamia. When analyzing the related stories, one is led to the conclusion that they describe transhumance. Laban, Jacob's uncle, was close enough to the well[5] for Rachel to run to her father and tell him the news about Jacob's arrival. But when necessary, Laban kept his herds as far as three days journey away from home (Gen 30:36). Jacob could not have escaped if Laban had stayed home rather than going away to shear his flocks (Gen 31:19), and it took three days to reach Laban and alert him to Jacob's escape (Gen 31:22). On the way to Canaan, Jacob and his family lived in tents (Gen 31:33–34), but upon his arrival in Sukkot, he built himself a house[6] and shelters for the herds (Gen 33:17). After arriving in Canaan and settling near Shechem, Jacob purchased land and built an altar (Gen 33:18–19), thus establishing a homebase.

From this point, Jacob's life mirrors the transhumant mode previously described in Padan Aram. This is suggested by the story of the Rape of Dinah (Gen 34). Information gleaned from the story reveals that since his sons returned home soon after the rape took place, they were tending flocks not far away. Jacob's life of transhumance is further described by other episodes related to him. Although Jacob continued moving, his wandering was limited; and it appears that he moved only to get back to his father's place of habitation. Once he arrived there, he settled in Hebron and made that his home base; as is apparent from the Joseph story (Gen 37:12–14), Shechem became a point in the circuit to which his sons kept returning. It seems that their seasonal return to the area made them well-known to the local inhabitants, because otherwise how could a total stranger help Joseph find his brothers (Gen 37:15–17)?

Nomadic herds and their keepers follow seasonal vegetation growth and migrate from region to region as dictated by grazing conditions. They do not have a home base or any form of permanent shelter, but live in tents. During the nomadic cycle, they cover distances longer than those of any other production system. For logistical reasons, the social unit of herders—the tribe or family—may have more than one flock with which they move from place to place. Herds can be very large, from 150,000 to 200,000 animals,[7] and they are comprised of one species or are mixed (Yalçin 1986:12–3).[8] When examining biblical herding practices, scholars use ethnographic descriptions of Bedouin groups in the Negev and Sinai for comparison because of their geographical proximity. Wandering is a well-organized undertaking that is based on experience gained through years of practice. Its organization and precision resemble a military undertaking (Ginat 1966:44), and the circuit includes visits to holy sites (Levi 1978:274), cemeteries, and celebrations of certain feasts.

While the historicity and date of the Patriarchal period are still being debated, the related biblical stories make it possible to reconstruct the lifestyle of those engaged in herding in Canaan. The survival of these traditions in the biblical narrative was made possible because of the familiarity of the writers/ compilers with the described practices. The pastoral nature of the early Isra- elites is partially demonstrated by the names Rebecca (*ribqâh* 'a row of tied animals') and Rachel (*rāḥēl* 'ewe'), both of which originated in pastoral ter- minology. The latter name might be related to the actual wandering, known in Arabic as *raḥil* (also *riḥleh* 'wandering, moving').[9]

That the Israelites were familiar with the nomadic system is suggested by Abraham, whose mode of life is a good example of that of a nomadic herder. Upon his arrival in Canaan, Abraham traveled throughout the hill country (Gen 12:6, 8) and even after pitching his tent between Bethel and Ai, he continued to wander southward (Gen 12:9). Upon returning from Egypt, where he stayed because of a drought in Canaan, he continued to wander until he came to Hebron and pitched his tent at Elonei Mamre (Gen 13:17– 18). Immediately after his return from Egypt, Abraham discovered that his flocks were too numerous to share grazing lands with those belonging to his nephew, Lot, so the latter chose the area around Sodom for his wanderings (Gen 13:12).[10] Abraham continued to live in a tent (Gen 18), and to wander as far as the Negev and Gerar (Gen 20).

Abraham's progeny show differing herding patterns. Isaac continued to live as a nomadic herder, residing in a tent he inherited from his parents (Gen 24:67) and wandering in the same circuit as his father. The principal reasons for maintaining the circuit were scarce water sources and the rights to use them (Gen 26). As described above, however, Jacob broke with tradi- tion upon his return from Padan Aram, where he had learned how to be a transhumant herder. Historically speaking, it is possible that this change was forced on him because of the new political situation that resulted from the rise of urbanism and the formation of strong and fortified cities which con- trolled their environs.

The nomadic system was not limited to the time of the Patriarchs and it was practiced even to the end of the Monarchy. One good example of a group that maintained this way of life is the Rechabites, who were told by their ancestor, Jonadab, son of Rechab: "You shall never drink wine, neither you nor your children; nor shall you ever build a house, or sow seed; nor shall you plant a vineyard, or even own one; but you shall live in tents all your days in the land where you reside" (Jer 35:6–7). Although herding is not specifically mentioned here as their means of livelihood, it can be safely presumed.

Caring for the Animals
Taking care of animals includes two main activities—sheltering and feeding—and each of the different production systems responded to these needs in its own particular way.

Sheltering
Nomads house their herds near their tents either in enclosures made of thorny plants that are piled up to form circular walls or in natural caves (Epstein 1985:43). This method is used also in the transhumant system during wandering periods. In biblical times similar practices were followed, as evident from the biblical text where the terms *gidrôt ṣo'n* (Num 32:16, 36) or *miklâh ṣo'n* (Hab 3:17) are used for animal enclosures. In some instances, as in 1 Samuel 24:4, a protective wall was placed near a cave that must have been used for sheltering the flock.

When sedentary or transhumant herds are kept at a permanent location to which they return at the end of each day, they are housed in stone-walled pens attached to buildings or compounds, or on the ground floor of houses in the city. Remains of stone-walled animals pens were excavated at several sites in the hill country and the Negev.[11] Scholars suggest that sheltering the animals inside the house was common among the Israelites who lived in "four-room houses" (Netzer 1992:196–8). Evidence from recent excavations strongly suggests that the four-room houses had more than one story, thus it has been suggested that the ground-floor space was used for storage, domestic activities such as milling and food preparation, and the sheltering of animals. Furthermore, it has been proposed that sheltering animals in this space helped keep the house warm during cold winter nights.

Protecting herds when they are grazing is highly important. Modern Bedouin consider the wolf and the hyena the greatest enemies of their herd (Abu-Rbei'a 1990:12). Attacks by wolves have even been documented recently (Pravolotsky and Pravolotsky 1979:57). According to one story, David told Saul that he rescued his father's herd from attacks by lion and bear (1 Sam 17:34). The prophet Amos, who was a *bôqēr* 'herder' (Amos 8:14–15), was clearly familiar with such situations when he used them as a metaphor for predicting the fall of Samaria and the Northern Kingdom: "As a shepherd rescues from the jaws of a lion a pair of shin-bones or the tip of an ear,[12] so will the Israelites who live in Samaria be rescued" (Amos 3:12). Both today and in the biblical period, dogs are used both for guarding the herd against predators and for keeping it together (*kalbēy ṣo'nî*, Job 30:1).

Feeding

The typical annual pattern of herd feeding includes grazing on green plants during winter (December–April), grazing on stubble and withered grass in spring and summer (May–July), and adding supplements of hay and grain to the withered grass in summer and fall (August–November). During the milking season (December/January–June/August), grain is given at night to entice the animals to return home where they can be milked. Feeding practices vary depending both on locations and climatic conditions affecting vegetation and on the mode of life practiced by a particular group.

Grazing is the most commonly used feeding method, but it has certain aspects that herders must consider. While these considerations may not be recognized and practiced continuously, but the way each of the production systems developed suggests that they are known and that their positive or negative influences are appreciated. Herders try to reach optimum conditions while avoiding overgrazing; thus they will limit the size of the herd to the carrying capacity of the land. Israelite familiarity with this problem is reflected in Genesis 13:5–7, where Abraham and Lot have large herds and "the land could not support both of them living together." Present day herding societies such as the Sinai Bedouin observe rules that help guide the behavior during grazing. These rules (in Arabic, *'anwa'ah*) place restrictions on grazing during certain periods in certain location. Transgressors are fined very heavily (Pravolotsky and Pravolotsky 1979:62–3).

Another determinant for grazing is whether agriculturists till the land as opposed to using it for foraging, or let the herds feed on the stubble rather than burning it. Since fertilizing with dung is one of the benefits gained by letting herds graze on the fields, the determination is usually in favor of allowing the herds on the fields. This is economical and beneficial to the farmer because while the stubble and weeds are being removed, the fields are fertilized and useless flora is converted into meat and milk. This can take place with herds belonging either to the landowner or to herders who pay for the grazing privilege. Both ways, the landowner gains.

Watering

Water resources are another factor that affects how animals are cared for. Herding cannot exist without an ample supply of water. Herds are watered daily during the year, and under the best conditions, in the summer heat the animals are given water twice a day. Watering is normally done along the daily route when the shepherd leads the herd to a water source. Animals usually get water at the beginning and the end of the day; and if there is no water source along the route, the herd is watered at home before going out and upon its return (Pravolotsky and Pravolotsky 1979:59–60). Water is drawn from a water source, such as a well or cistern, and poured into troughs (*rĕhāṭîm* 'trough,' Gen 30:38, 41; *šoqet* 'troughs,' Gen 24:18; *šiqatôt hammayim* 'watering troughs,' Gen 30:38) that, in many cases were made of carved stone. Watering the herds is described vividly in the episode recounting Jacob's first encounter with Rachel (Gen 29:2–10).

Questions of water rights dominate many of the biblical stories depicting herding. The dispute between Abraham and Abimelek over a well (Gen 21:22–34) is repeated when Isaac's life is described; however, in Isaac's case the dispute was over several wells which were dug by him and filled by his adversaries (Gen 26:15–33).

The well was always the hub of life of the herding community; it was not just a source for water, but also a copious source of information. It is the background for several biblical episodes in which people meet by the well to procure information. Abraham's servant, who was looking for a wife for Isaac, came to the well to conduct a test and found Rebecca there (Gen 24:11–22, 42–46). When Jacob escaped the wrath of Esau, he came to a well where he inquired into the whereabouts of his relatives (Gen 29:2–10). Moses, on his flight from Pharaoh's judgment, met Jethro's daughters by the well and found shelter in Midian (Exod 2:15–21). All of these stories mirror a situation encountered daily by many Israelites in the villages and towns, or on the road.

The Shepherd

The shepherd (*rô'eh*) was the one who cared for the animals on a daily basis. A good shepherd takes care of the flock, leads it to pasture and water, and protects it against any mishap (Ps 23:1–4), but who within a community was a shepherd?

A comparison of biblical references and ethnographic studies conducted among preindustrial Near Eastern societies enables us to put together a profile of the shepherd. Among the Bedouin, a young girl of eight to ten years old first starts taking the herd out to pasture as a trainee and later as an independent shepherdess. She continues in this role until the age of fifteen or sixteen, when most girls get married and then stay home (Pravolotsky and Pravolotsky 1979:69). Although sometimes young boys are found in the role of the shepherd, it is mostly reserved for girls. A similar picture is depicted in the biblical stories concerning Rebecca, Rachel and Leah, and Jethro's daughters, but there are quite a few passages where men, and not necessarily young men, were shepherds.

Being a shepherd was not denigrating. David, son of Jesse, who was the youngest among his brothers, was supposedly looking after the herds while he was elected future king of Israel (1 Sam 16:11). He continued to look after the herds when his older brothers joined Saul's forces against the Philistines possibly because he was too young for military service (1 Sam 17:14–15).

Most of the information concerning adult male shepherds, however, appears in cases where conflicts arise. The stories relating the disputes between the shepherds of Abraham and Isaac and those of Abimelek, king of Gerar, have been already mentioned, but should be considered again here because they depict fighting adult males on both sides. The same might be true in the case of the male shepherds who chased Jethro's daughters away from the well. "And when they filled the troughs to water their father's sheep, some shepherds came and drove them away. But Moses came to the help of the girls and watered the sheep" (Exod 2:16–17). Both Jacob and Moses were adult shepherds in the service of their fathers-in-law. Their mature age is indicated by the fact that they were ready to marry once they settled at the place to which they escaped.

Jacob's sons, who were with the herds when their sister Dinah was raped, were old enough to avenge her honor (Gen 34:5, 25). Joseph, at the age of seventeen, an age ripe enough for marriage in these communities, was herding with his older brothers when the pasture was close to the home base. But when the herds went on the *riḥleh*, he did not participate (Gen 37:2, 12). By that time, Judah who participated in caring for the herds (Gen 37:26) was old enough to get married and father children (Gen 38). Amos the prophet was involved in sheep herding, but it seems that his involvement was on a more professional level, since he was not just a *rôʾeh* (shepherd) but a *nôqēd* (Amos 1:1).[13] Another famous *nôqēd* was Mesha, king of Moab (2 Kgs 3:4). Using this term in reference to a king suggests that it was reserved for someone who was more than just a simple shepherd, but rather to a sheep breeder.

Not all shepherds own their flocks or are members of the immediate family that owns the herds. Hired shepherds who have a contract between themselves and the owners are well-known in preindustrial societies. In such instances, the shepherd receives as a salary, material for clothing, one pair of sandals per year, a daily food allotment, and one kid per year from each owner (Pravolotsky and Pravolotsky 1979:71). An example of such a contract is the one between Jacob and Laban, in which Jacob at first worked for Laban in return for marrying his daughters (Gen 29), and later in return for all of Laban's spotted animals (Gen 30:32–34).

Being a shepherd is a way of life with certain daily activities. Women tend to use the time for spinning thread, so they carry a large amount of raw wool and a drop spindle with which to make yarn. To entertain themselves, shepherds carry wind and string instruments. Is it possible that David's ability to play a musical instrument (1 Sam 16:23) was enhanced during the long hours of watching the herd with nothing better to do?

An important part of the shepherd's duties is to protect the herd from thieves and animals of prey (Ezek 34:5–10). For this reason, the shepherd carries a stick and learns to throw stones with the sling.[14] According to biblical traditions, David was a shepherd capable of protecting his herd. Although it was only according to his own testimony, he protected his herd against a lion and a bear (1 Sam 17:34–36), possibly using defensive skills with a stick and sling. According to the biblical story, David used all these skills he supposedly perfected as a shepherd in his duel with Goliath. Being a good shot, he hit Goliath in the forehead (1 Sam 17:49). The shepherd's great responsibilities gave rise to metaphors that compare the leaders of the nation of Israel to shepherds (Jer 23:1–2; Ezek 34) and name YHWH as the shepherd of his people (Ps 23).

That the shepherd's life was hard can be illustrated by Jacob's speech of complaint against the treatment he received from his father-in-law, Laban:

> *In all the twenty years I have been with you, your ewes and*
> *she-goats have never miscarried. I have never eaten rams*
> *from your flocks. I have never brought to you the carcass*
> *of any animal mangled by wild beasts, but I bore the loss*
> *myself. You demanded that I should pay compensation for*
> *anything stolen by day or by night. This was the way of it:*
> *the heat wore me down by day and the frost by night; I got*
> *no sleep. . . . I worked fourteen years for you to win your*
> *two daughters and six years for your flocks, and you*
> *changed my wages ten times over.* (Gen 31:38–41)

Breeding

There is no question that, in the Iron Age (and much before), people knew about breeding to improve the stock or produce new breeds. The best example of the practice was the breeding of mules.[15] The Israelites knew very well that two different species cannot be crossbred, so the interpretation of the allusion to plowing with an ox and donkey together (Deut 22:10) as a reference to cross-breeding is incorrect. However, a very good example of breeding for stock improvement or the enhancement of certain traits is found in the story of Laban and Jacob (Gen 30:32–31:12). When Jacob was asked what he would want as his salary for tending Laban's flocks, he responded by suggesting that all the spotted, brindled or brown sheep be given to him. Knowing that the majority of the sheep were white, Jacob expected Laban to agree to his suggestion. However, after the agreement was reached, Jacob decided to influence the outcome, and when the sheep were in heat he allowed them to be impregnated while facing wooden rods he had peeled to create a spotted effect. This follows the superstition that a pregnancy can be affected by what the pregnant female sees. However, Jacob did not trust this superstition alone, and he only allowed spotted rams to mount the strong females in heat (Gen 30:40–41). "So Jacob's wealth increased more and more until he possessed great flocks" (Gen 30:43). In modern terminology, it would be considered genetic engineering.

SMALL CATTLE

Small cattle—sheep and goats—were the most common domesticated animals during the Iron Age. The prevalence of small cattle is evidenced linguistically in biblical and extra-biblical sources as well as supported by zooarchaeological finds at many sites. Biblical references to small cattle (*so'n, miqneh*) are numerous and include references to the two main components of the group: sheep (*kebeś*) and goat (*ʿēz*). Furthermore, in the ancient Near East, riches were in part measured by the size of the herd (*ʿēder*) a person owned. Sinuhe, the Egyptian nobleman exiled to Canaan during the mid-twentieth century B.C.E., describes his wealth in terms of the numbers of his cattle saying "I became extensive in my wealth, I became abundant in my cattle" (Pritchard 1969a:20-2).

Environmental conditions play an important role in raising small cattle, because differences in precipitation and geological formations in diverse geographical regions to a large extent determine the makeup of the herds and the culture and beliefs of the communities that own them. In Palestine, conditions in the south are much different than in the north, and there are differences between the valleys and the hill country. Since goats are more adapted to harsh conditions, they would make up the majority of the herd where unfavorable conditions exist.

The place of small cattle in the economy of ancient Palestine can be illustrated by information gathered in recent preindustrial times. In the last quarter of the nineteenth century, small cattle comprised 25% of the total income in Palestine, while orchards were 15% and field crops 60% (Avitsur 1972:222). Ottoman records show that the number of small cattle at the beginning of the twentieth century was a quarter million (Avitsur 1972:223), with 180,000 in the Jerusalem region alone (Ilan 1984:65). In 1920 the British Mandatorial Government estimated that there were 205,967 sheep and 325,512 goats in Palestine (Ilan 1984:65). The first census of small cattle taken by the British Mandatorial Government in 1930 counted 252,773 sheep and 440,132 goats (Hirsch 1933:5), out of which only 6991, or 1%, were owned by Jews. In the same year, it recorded that income from one sheep was 600 mils compared with 290 mils per one dunam (1/4 acre) of field crops (Avitsur 1972:223). The income from forty-eight sheep was equal to that of 100 dunams (25 acres) of field crops. In the nineteenth century, the price of sheep was higher than that of grain, and this explains why, according to Conder of the Palestine Exploration Fund, the village of Yatta, south of Hebron, had 1,700 head of small cattle, out of which 250 belonged to the *mukhtar*, the village head (Ilan 1984:67). These data from preindustrial times, which include comparative prices and animal population estimates, illustrate that herding then, and most likely also in ancient times, was a profitable occupation and economically advantageous.

Products from Small Cattle
Sheep and goats are raised for wool or hair, milk and its products, meat, skins, and dung. Common products of sheep and goats are discussed here, others—such as hair, wool, and skins—are described in later chapters.

Fresh Milk
The Story of Sinuhe (ca. 20th–19th century B.C.E.) describes the land of Yaa (Canaan) as a place where "there was no limit to any [kind of] cattle" and he was offered "milk in every [kind of] cooking" (Pritchard 1969a:19–20). Thirst was quenched by drinking preboiled milk; as described by Sinuhe, "The sheikh...gave me water while he boiled milk for me" (Pritchard 1969a:19).[16]

The significance of milk (*ḥālāb*) in the biblical diet is apparent since it connotes plenty and nutritional richness when mentioned in association with another whole and complete food, honey (*děbaš*). The wholesomeness of these two foods was not lost on the biblical writers and was widely understood in biblical circles, from which the expression emerged describing the Land of Israel as "flowing with milk and honey" (Exod 3:8; and elsewhere). This combination was also used in many ancient societies for sacrifice (Brothwell and Brothwell 1969:73).

Sheep and goats start producing milk as soon as they give birth (December–January)[17] and continue through the summer (June–August) (Fig. 2.2). After weaning (March–April),[18] milking becomes a daily chore that is done twice a day, in the morning and evening, before and after grazing. A sheep's daily average may vary from ¼ liter to 1¼, liters with an average

Figure 2.2. *Black goats with a mother and kid in the front.*
(Photograph by the author.)

individual yield per season of 80 liters or more (Hirsch 1933:29). Daily milk
production of a goat is 1 to 2 ½ liters during the lactation period, which lasts
seven to eight months (Hirsch 1933:57).

Milking in preindustrial societies was done by setting the ewes and does
one opposite the other and tying them in pairs by a long rope, creating two long
rows (Hirsch 1933:28, 60) (Fig. 2.3). In biblical terminology, a row of tied
animal was termed *ribqâh*,[19] the name given to Isaac's wife, Rebecca. A cylinder seal from Mesopotamia depicts this process, showing one shepherd holding
a long-thong whip while another milks the goats. In the upper part of the seal,
small disks seem to represent cheese drying out (Aynard 1972:52). Another seal
impression, dated to the Early Dynastic III (middle of the third millennium
B.C.E.), depicts a man milking a goat while seated on a stool inside a stable; a
guard dog lies next to the stool, and vessels, possibly containing milk products,
are arranged in a row above the man and the goat (Pritchard 1969b:fig 97).

Figure 2.3. Ribqâh; *a group of sheep tied together for milking.*
(Photograph by the author.)

Milk has always been a very important dietary component in the Near
East, but since it cannot be preserved for a long period, only a little is used in
its fresh state and the rest is turned into several other products. Depending on
the size of the herd and the quantity of milk produced, some of these products
could be marketed beyond home use.

Processed Milk Products

Dairy products constituted an important category of food in ancient Israel.
Whenever possible, milk was drunk to quench thirst (Judg 5:25), and from
the number of times milk is mentioned and herds of sheep and goats appear
in the biblical text and in extra-biblical records related to this region, we can
assume that it was available in abundance for consumption and processing.
One often-mentioned dairy product is *ḥemʾâh*, most likely 'curds' or 'butter'
(in Arabic, *zibdeh*). It was served in a *sēpel* (Judg 5:25), probably a bowl, and
one way of eating it was with honey (2 Sam 17:29; Isa 7:15, 22; Job 20:17).
Another dairy product was *gĕbînâh* (Job 10:10), probably cheese (in Arabic,

labaneh). A different kind of cheese, is *ḥarîṣ ḥālāb* (1 Sam 17:18; in Arabic, *salfiḥ* or *ʿafig*),[20] mentioned among the presents sent by Jesse to the military commander in charge of David's brothers. Another certain dairy product, *šĕpôt bāqār* (2 Sam 17:29), which was served to David and his people when he escaped from his son Absalom and went to Mahanayim, could have been the Arabic *semneh* (see below).

Processing dairy products makes milk available even to some of those people who cannot drink it fresh. As Davis notes, "by the process of converting the lactose of fresh milk to lactic acid, as happens when yoghurt is prepared (using bacterial cultures), or further removing the lactic acid to produce cheese, these milk products become digestible even to lactase-deficient" (1987:156).

Some of the dairy products mentioned in the Bible are still being made in preindustrial societies, most likely following the same methods.[21] Since there are some variations in how these products are made, the following is a general description based on information from several sources relating Near Eastern practices.

- *Lebben* (yoghurt). This important summer delicacy is prepared by pouring the fresh milk into a goatskin container[22] and adding some buttermilk. The container is left hanging outside overnight, during which time the milk turns sour and curdles (Hirsch 1933:38).[23] *Lebben* can be drunk in cups and bowls or used for dipping bread. Avitsur suggests that this might be the *ḥomes* mentioned in Ruth 2:14 (Avitsur 1976:64).
- *Zibdeh* (butter; *ḥemʾah*). Fresh butter is made by processing the *lebben*. Churning starts in the morning, when the air is still cool. The skin is blown up with air, hung from a tripod or a branch, and swung back and forth for about one and a half hours. This separates the fat particles, and the upper layer becomes butter that can be used in cooking or for spreading on bread (Hirsch 1933:38; Pravolotsky and Pravolotsky 1979:64).
- *Semneh* (melted butter). When there is a sufficient amount of *zibdeh*, *semneh* can be made. Otherwise, the *zibdeh* is stored in a vessel with salt for future processing. To make *semneh*, the *zibdeh* is cooked in a pot (preferably copper) on an open fire with a mixture of salt, grain

or flour, saffron, and other herbs. The contents are brought to a boil and stirred constantly while scum and dirt are skimmed off. When the fat becomes clear, after about one and a half hours, the fire is lowered until the fat is completely clear and the other components settle at the bottom. The fat is processed through a filtering cloth into a vessel that, after cooling, is closed. The leftover liquid can be further processed to make *kishik* (see below). The *semneh* can be transferred to a goatskin for up to several years' storage (Hirsch 1933:39; Pravolotsky and Pravolotsky 1979:64). This dairy product is possibly *šĕpôt bāqār* of 2 Samuel 17:29.

- *Kishik* (dry cheese). This type of cheese, possibly the biblical *ḥarîṣ ḥālāb*, is made out of the buttermilk[24] left over from making *semneh*. It is mixed with salt and placed in a cloth bag to dry. After two or three days, when the water has evaporated, the mixture is taken out and formed into balls or lumps that are put in the sun to dry.
- *Labneh* (cheese; *gĕbînâh*). The exact origin of cheese is unknown. It was probably discovered by accident when a freshly harvested stomach was used for storing milk. The admixture of two clotting agents, rennin and lactic acid bacteria, created a cheese-like product (Ishak 1987:221).

  Labneh, the most typical cheese of the Near East, is an acidified soft cheese made of cow, goat, or sheep milk, in a mixture or separately.[25] The milk used for the process can come from evening or morning milking, and can be fresh or boiled. It is poured into a container with some lactic bacteria (acid milk, coagulated milk, or yoghurt) or a small piece of kid's or lamb's rennet[26] for clotting. After the milk has curdled, it is placed in a cloth bag for draining. Following that, salt is spread on the surface to prevent microbial contamination and improve flavor. *Labneh* is moist and sometimes has a very thin crust. It contains 20–30% dry matter and a fat content of 45–50% (Ishak 1987:230). It can be pressed into molds and is eaten fresh or kept for a while in olive oil. For long-term preservation, up to 5% salt is mixed with the curd, which is then formed into balls that are placed in the open air to dry. This product is stable for several months (Hirsch 1933:40; Ishak 1987:231).

Meat

Sheep and goats were raised for their meat (*bâśār*), which was prepared in several ways and eaten on special occasions. Meat was not consumed on a daily basis for several reasons. Because killing an animal terminates its productivity, the slaughter of any animal had to be carefully calculated. Lack of means for preservation dictated that the meat had to be consumed immediately after slaughter.[27] Davis maintains that, for the sake of efficiency, "in a meat-producing economy it makes sense to slaughter near the end of an animal's immaturity . . . " (1987:157). There are two optimal ages for slaughtering lambs and kids, immediately after weaning[28] and at the age of one year (Num 6:14; 7:15).[29] Slaughtering or selling a recently-weaned lamb (*śeh,* Exod 12:5) or kid (*gĕdî ʿizzîm,* Judg 6:19; *śĕʿîr ʿizzîm,* Lev 9:3) is most economical because at this time the ewe is at the height of milk production and no major investment in the young animal, such as feeding, has been made by the herd owner. When Samson came to visit his wife and brought her a kid as a present (Judg 15:1), the animal must have been recently weaned because the event occurred at the height of wheat harvest, in April or May (Borowski 1987:32).[30] Keeping an animal for one year gets it to the point of reaching maximum meat weight with minimum investment. This must be the reason for the prescription of offering yearlings for sacrifices, which is mentioned twenty-four times in Numbers 7.[31] This cultic precept is probably depicted in a free-standing statue from Mari (18th century B.C.E.) of a bearded man holding a lamb (a yearling?) in his arms that is possibly intended for sacrifice (Davidson 1962:165).

In some cases, animals between the ages of one and three are considered to be at the optimal meat-production stage relative to investment in feeding (Wapnish 1993:430). From an economic point of view, the herder needs to cull the flock. He cannot afford to have too many unproductive males that only consume food, and the number of females has to be restricted to the carrying capacity of the land. Therefore, males not destined to become studs are slaughtered early; female milking goats (does) are usually slaughtered when they are seven or eight years old (Hirsch 1933:61), when they become less productive and newly matured females replace the slaughtered ones.

In antiquity, meat and milk production were very important endeavors in which private individuals, as well as temples and the royal palace, participated.

For example, it is mentioned that King David had large herds of cattle in the Sharon Plain and in the valleys, and flocks of small cattle at undisclosed locations (1 Chr 27:29, 31). Uzziah "erected other towers in the wilderness and dug many cisterns, for he had large herds of cattle both in the Shephelah and in the plains . . . " (2 Chr 26:10). The Levites were given large tracts of land (*migrāš*), where they were supposed to "keep their animals, their herds, and all their livestock . . . " (Num 35:3; see also Jos 14:4).

A small number of animals was kept for fattening. To increase their weight, these animals were tied to restrict their movement and were fed a special mixture. Fattened sheep, *mēḥîm* (Isa 5:17),[32] were considered choice sacrifice (Ps 66:15) and the term was used metaphorically to describe the rich. Fattened rams, *karîm* (1 Sam 15:9; 2 Kgs 3:4),[33] were also considered a delicacy (Ezek 39:18; Amos 6:4). It seems that raising and selling fattened animals were lucrative businesses in which kings (2 Kgs 3:4) and other leaders (Ezek 27:21) were involved.[34]

Fat, *ḥēleb* (Ezek 34:3), was considered a nutritious food element, therefore, from very early times (Gen 4:4) it was a major component of sacrifices (Exod 29:13). Because of its importance, *ḥēleb* was used metaphorically when referring to choice produce of the land (Ps 81:17; 147:14).

When an animal was selected to be slaughtered, there were several ways of preparing the meat.[35] Meat could be cooked or boiled in water (Exod 12:9),[36] but there was a strict prohibition against cooking a kid in its mother's milk (Exod 23:19; 34:26; Deut 14:21).[37] A by-product of boiling meat was broth (*māraq*, Judg 6:19–20; Isa 65:4). Although meat was roasted on a fire (*ṣĕlî ʾēš*, Exod 12:8–9), we do not know how common this was, since it is mentioned very few times, twice in connection with the Passover sacrifice. However, its mention in Isaiah 44:19 suggests that roasted meat was part of the daily diet.

Dung

Dung, *gālāl* (1 Kgs 14:10) is another by-product for which animals were raised.[38] Dung can be collected in animal pens and stored in piles for future use, or it can be spread naturally on harvested fields during grazing (*domen*, 2 Kgs 9:37). As Krispill notes, when animals are kept in caves together with the family, daily clearing of dung is essential and has to be done before fermentation begins (1986:27). Dry camel, equid, and cattle dung is used as fuel, while small cattle dung is usually not used for this purpose.[39]

Bones and Horns
Most of the bones found in archaeological excavations belong to animals that were consumed as food,[40] but a small number of bones recovered have been worked. Worked bones of ruminants were used in many ways, shapes and forms for a variety of purposes. The most common skeletal elements used were long-bone shafts, rib bones, scapulas (shoulder bones), astragali (knuckle bones), and sometimes skull bones (Kolska-Horwitz 1990:145). The type of bone used was mostly dictated by the type of artifact produced. Astragali were used as game pieces, while ribs were usually used for making spatulas, long bones were shaped into awls, points, buttons, and inlay pieces. Other objects made of bone include the so-called "fan handles" (Ariel 1990:134–6), pendants, beads, spindles, spindle whorls, rods, combs, knife handles, and more. Long bones of large ruminants (cow, for example) could be used for making flute-like musical instruments.[41]

Horns were mostly used as tapping (hammering) tools, especially in flint tool-making, and as musical instruments. The latter function is well documented in the Bible. A horn used as a musical instrument is known as *šôpār*, and was used in times of peace in the cult centers (Ps 98:6), for warning or assembling the people (Judg 3:27), and in war (Judg 7:22).

Zooarchaeological Remains
Sheep- and goat-raising during the Iron Age is well-established in the archaeological record; there is hardly an archaeological site where remains of small cattle are not available in large quantities. While an enumeration of sites yielding such remains is thus pointless, there are definite issues that should be raised. The anatomical closeness between sheep and goats sometimes does not allow a precise identification of the species, and the results are presented as "sheep/goat." Nevertheless, identification of the age and sex of the animals enables a study of the local economy to be made. The presence of older specimens, three-year-olds and older, indicates that the animals were kept as long as possible for their by-products, while remains of younger specimens, up to three years old, indicates a meat economy in which either the

animals were slaughtered where they were raised or were brought in for consumption from elsewhere. The recovery of certain bones at particular locations on a site can help reconstruct the social structure of a community in which certain segments of society could afford the choice parts of an animal, such as the breast or the front thigh. Cut marks on the bones indicate whether the butchering was done by an experienced person or a novice, and they can indicate how the parts were used when prepared for consumption. Evidence of burning also points out how the parts were prepared.

The number, weight, and type of the recovered bones can help in computing the number of animals present at the site.[42] This, in turn, can help in computing the amount of meat, wool, milk, and other by-products that was available for the local economy.

Zooarchaeological examination of animal bones to determine the age of the animals at death yields some insights into historical processes, social as well as economic, that can be correlated with written evidence. Results of osteological studies of material from Tel Dan "are consistent with theories that identify the settlers in the earliest Iron Age communities as sheep and goat agropastoralists" (Wapnish 1993:434). This conclusion is based on the fact that the percentage of large cattle bones in the Late Bronze Age was greater (half of the slaughtered animals) than during the twelfth to tenth centuries B.C.E. (one quarter of the sample). Furthermore, as time went on, the percentage of sheep/goat bones decreased to 51%, while cattle bones increased to 49%. However, there were also changes in the mortality age of the sheep and goats; more animals were killed in their second year than in their first. "This change suggests a reorientation of sheep and goat husbandry towards meat production" (Wapnish 1993:430–31). Wapnish proposes that these changes can be correlated with historical events recorded in the Bible. The changes in the osteological record (LBA to IA) suggest a cultural change such as the takeover of Laish/Dan by the Danites in the early Iron Age (Judg 18). As the settlers became urbanized, another cultural change took place vis-a-vis the use of animals, and this change is also reflected in the osteological record.

The analysis of sheep/goat bones from seventh-century B.C.E. layers at Tell Jemmeh and Tel Miqne-Ekron reveals that the majority belong to sheep over three years old and demonstrates a major economic shift in both sites.

This suggests "that the shape of the animal economy resulted from special demands on the pastoral producers. Assyrian markets and Assyrian tribute demands may have been too attractive to permit the local consumption of highly marketable resources" (Wapnish 1993:439). This conclusion is based on the complete absence of bones of the marketable stock, the zero to three year old animals. It appears that the local herders preferred to market the young animals, and their own diet included only old animals. However, since the economy of both sites did not collapse, it seems that the process was well managed. That the Assyrians exported large numbers of animals as booty and tribute is well attested in their records. For example, Ashurnasipal II (883–859 B.C.E.) reports that, on his expedition to Carchemish and the Lebanon, he received at one point "1,000 (head of big) cattle, 10,000 sheep . . ." (Pritchard 1969a:276).

Goats

According to Davis, the goat (*Capra aegagrus*) may well be the first domes-ticated animal in the Near East (1987:140) (Fig. 2.4).[43] There are two main types of goats in the region: the Bedouin, or black, goat is raised mostly in the Negev and Sinai, and the hill country, or Damascene, goat is raised near per-manent settlements in the Judean hills, Samaria and Galilee. A third type of goat, Baalbek, with black and white hair, is raised in the Golan and Hermon mountains. The Bedouin goat has black hair, short ears, and brown eyes; it weighs 12–25 kg, and produces an average of 75 liters milk per year. The Damascene goat has reddish-brown hair, long ears, blue eyes, weighs 30–50 kg, and produces about 400 liters milk per year. The hill-country goat, like the Awassi sheep, needs to drink twice a day, while the Bedouin goat can be without water for as long as four days.[44] This trait enables the Bedouin goat to survive and flourish in arid and semiarid areas where the other small cattle cannot, thus an area with a carrying capacity of 64 Bedouin goats (about 960 kg) might not be able to support even one sheep or one hill-country goat for lack of water (Shkolnik 1977:100–104). While the black goat is fed mostly by grazing, the Damascene goat is kept near the home, and is fed on greens and feed.

Keeping goats is very profitable because they are not selective in their diet, they mature quickly, and they are a good source of milk, meat, and hair

Figure 2.4. *Black goat.* (From Bodenheimer 1935:pl. XII.)

(Ilan 1984:67). Nevertheless, their nonselective diet causes serious environmental damage, and this has always been a source of debate whether the goat should be permitted within certain regions.[45] In 1950, the State of Israel legislated restrictions on goat herding, allowing grazing only on private land at a rate of one goat per ten dunams of irrigated land or one goat per forty dunams of unirrigated land. The Mishna and the Talmud also restrict areas for grazing goats; but since their concerns are mostly with damage to agricultural land and produce, they permit grazing goats in certain types of forests (Kaplan 1989:27–28). The argument concerning the black goat surfaced again in Israel in 1978, culminating with recommendations of controlled grazing (Baily 1978). That goats were destructive must have been well known in the Ancient Near East. A free-standing statue of a goat leaning its forelegs on a tree (as if ready to chew on it), made of gold foil, shell, and lapis over a wood core, was found at the Royal Cemetery of Ur (ca. 2500 B.C.E.) (Davidson 1962:143; Pritchard 1969b:figs 667–68).

Although goats are hardier than sheep, there are certain diseases that are common to both species, especially anthrax, brucellosis, contagious ecthyma, and foot-and-mouth disease. Goats are also afflicted by ticks and lice.

Several terms pertaining to goats are recorded in the Bible. A female goat (doe) is *ʿēz* (Gen 15:9); a male goat, *tayiš* (Prov 30:31) and *ʿattûd* (Gen 31:10). The Book of Daniel presents another term, *ṣāpîr* (Dan 8:21) or *ṣĕpîr*

ʿizzîm (Dan 8:5). A young he-goat is *śāʿir* (Lev 16:20) or *śĕʿîr ʿizzîm* (Lev 4:23); a young she-goat is *śĕʿîrat ʿizzîm* (Lev 4:28). Another term for a young he-goat is *gĕdî* (Judg 13:19) or *gĕdî ʿizzîm* (Gen 38:17). The term *śeh* (Exod 12:5) is sometimes used for a young he- or she-goat as well as for sheep. Artistic depictions also differentiate between sex and age of goats (Fig. 2.5). One example is that of the so-called Ur Standard which in the middle register depicts what appear to be a he- and she-goats (Davidson 1962:146–47). The he-goat has curved horns, and the genitals are well-defined; it is differentiated from the rams by its lack of wool. The she-goat can be differentiated from the adjacent rams by the "beard" on its chin.

Hair and Skins

Goats are shorn once a year, usually in May or June. Unlike sheep, goats are not washed before shearing because their hair is short and straight and generally clean. While the older goats have only their belly, chest, and flanks shorn, the younger ones have their entire body shorn. The hair from an adult animal weighs 300–500 gr, but the proceeds from the shearing only cover expenses (Hirsch 1933:60). The first to be shorn are goats that do not give milk, because the Bedouin believe that shearing lowers milk production (Pravolotsky and Pravolotsky 1979:67). In recent history, goat hair has been used mainly for making Bedouin tents and sacks for transport (Hirsch 1933:10).[46] It is highly possible that, in biblical times, goat hair was used for the same purposes or, at least, for rope making.

Figure 2.5. *The war and peace (reverse) Royal Standard from Ur depicting draft and herd animals.* (Frankfort 1954:36–7.)

Skins (*'ôr*, Lev 13:48–9) have been used as material for articles of clothing (Gen 3:21) and as material for writing. DNA studies of Dead Sea Scrolls parchment fragments indicate the wide use of goat skins for their production (Oppenheim 1996). Skins also had a major use as containers for liquids such as water, milk, and wine (*no'd*, 1 Sam 16:20; *ḥēmet*, Gen 21:14–5; *nēbel*, 1 Sam 1:24). Hirsch indicates that in the late 1920s, the exports of "waterskins" from Palestine were 58.4 tons, and this in addition to 1,410 tons of raw or dried skins of sheep, goats, and lambs (1933:11).

Processing skins is a lengthy and complicated process. The following is a description of skin processing in certain Near Eastern preindustrial societies. After slaughtering the goat, the skin is cut at a few choice places (around the hooves and neck) so it can be removed whole. If the skin is destined to be used as a container, special care has to be taken not to slice or puncture it. The skin is then buried for a week, after which time the hair comes off easily when rubbed with a stone. It is washed with water numerous times, and the leg openings are sewn. If it is to be used as a waterskin, it is filled with pomegranate rinds; if these are not available, a mixture of other plants can be used. This is done to cure the skin and neutralize the taste. If the skin will be used for holding milk, it is filled with a mixture of different plants. After a week, the skin is emptied and washed several times with clean water. Afterwards, the skin is filled for another week with black olives that were boiled in water and dried in the sun. It is again emptied and washed with water, and the processing is complete. When the skin is in use, the outer side of the skin is constantly wet because of its porosity, and this helps keep the contents cool (Pravolotsky and Pravolotsky 1979:68).

Ancient records and glyptic art make it very clear that inflated skins were used by soldiers for making rafts and life preservers when crossing rivers (Fig. 2.6). Ashurnasirpal II (883–859 B.C.E.) describes his expedition to Carchemish and the Lebanon saying: "I departed from the country Bit-Adini and crossed the Euphrates at the peak of its flood by means of (rafts made buoyant with inflated) goatskin (bottle)s" (Pritchard 1969a:275). This operation is shown in reliefs from his Northwest Palace in Nimrud, where one register describes soldiers blowing up skins while others swim across a river on skins (Yadin 1963:330, figs. 1–2). Another register depicts refugees escaping across the river floating on skins (Yadin 1963:330, fig. 8).[47] Shalmaneser III (858–824 B.C.E.) when reporting about his first year in his

Figure 2.6. *Fugitives crossing river with air-filled skins, from Nimrud.*
(Frankfort 1954: 85.)

so-called Monolith Inscription, also tells about a similar activity: "I crossed the Euphrates on rafts [made buoyant by means] of [inflated] goatskins" (Pritchard 1969a:277). During his sixth year, the Monolith Inscription again describes a river crossing he undertook: "I crossed the Euphrates another time at its flood on rafts [made buoyant by means] of [inflated] goatskins" (Pritchard 1969a:278).

Both, Ashurnasirpal II and Shalmaneser III boast of getting tribute of small cattle, which must have included goats (Pritchard 1969a:276–278). Other Assyrian kings who boast of receiving tribute and taking booty of small cattle in Palestine were Tiglath Pileser III (744–727 B.C.E.), Sennacherib (704–681 B.C.E.), Esarhadon (680–669 B.C.E.), and Ashurbanipal (688–633 B.C.E.).

In pre-Israelite times, the Egyptians also used Palestine as a source for animals. Thutmoses III (1490–1436 B.C.E.), after the battle at Megiddo, reports that the enemies brought "tribute of silver, gold, lapis lazuli, and turquoise, and carrying grain, wine, and large and small cattle for the army of his majesty . . . " (Pritchard 1969a:237), and among the booty he includes "1,929 cows, 2,000 goats, and 20,500 sheep" (Pritchard 1969a:237). He was followed by Amen-hotep II (1447–1421 B.C.E.) who, after the battle of Aphek in his ninth year, took as booty "its cattle, its horses, and all the small cattle (which) were before him" (Pritchard 1969a:246). He did the same at the battle of Anaharath, and on his return to Memphis he lists as booty "all (kinds of) cattle, without their limit" (Pritchard 1969a:247).

Sheep

The most numerous and wide-spread breed of sheep (*Ovis orientalis*) (Fig. 2.7) in southwest Asia, including Israel, is the Awassi. Its fat tail (*'alyâh,* Exod 29:22), the most characteristic attribute of this sheep, distinguishes it from other breeds (Epstein 1985:1–2). Artistic representations from Ur (Uruk III), Assyria, Arabia, and Israel show this tail and help determine that the sheep raised in antiquity were closely related to the Awassi sheep (Epstein 1985:2–6). Herodotus described the breed saying, "In Arabia there are remarkable sheep of which one breed possesses tails three ells long, so that each beast has its tail tied to a tiny cart in order not to drag it along the ground and wound it" (in Hirsch 1933). As Epstein observes, "such heavy fat tails severely impede locomotion and can only be developed in sheep protected by man" (1985:6). In some instances, such heavy tails prevent impregnation without human intervention.

The Awassi was the only sheep in preindustrial Palestine, and at that time the breed, including different sub-types that can still be found in certain areas from Iraq to Cyprus, could also be found in other regions of Asia and in Africa. The origin of the Awassi sheep is most likely in Iraq, where Bedouin legend places the Bedouin Awas tribe along the Euphrates (Hirsch 1933:14–5). In addition to the fat tail, this breed differs from the wild sheep (*Ovis ammon anatolica*),[48] in one important aspect, the type of coat. The wild sheep has a coat like a goat's, made up mostly of long hairs, between which a light,

Figure 2.7. *Fat-tail (Awassi) sheep.* (From Bodenheimer 1935:pl. XII.)

woolly undercoat can be found. A long period of planned breeding finally produced the prototype of our present sheep (Nissen 1988:24). However, the Awassi sheep is not primarily a wool producer, but is raised mostly for its flesh, fat, and milk (Hirsch 1933:19). The Standard from the Royal Cemetery of Ur depicts in the center of the middle register two sheep or rams appearing as part of a festive procession (Davidson 1962:146-7). They have thick, curved horns and their bodies are covered with wool that has the same texture as the hair of the goat right next to them. Other examples of early domesticated sheep in Mesopotamian art suggest that these animals were common. In Sumer, sheep are prevalent in all artistic modes. On an alabaster goblet from Uruk (3500–3000 B.C.E.) a ram is shown at the top register supporting a human figure that might represent the king (Davidson 1962:127). Sheep and rams are depicted on cylinder seals, in reliefs, and as figurines (Frankfort 1954:pls. 3c, 4a–b, 8a).

The Egyptians were also familiar with and respectful of the Ovid family. Egyptian art depicts sheep and rams in many ways. Minor Egyptian deities are shown having a ram's head, as in a papyrus illustrating the separation of the earth from the sky, where two ram-headed deities assist Shu, god of air, in this activity (Pritchard 1958:fig. 158; Davidson 1962:45). The god Khnum was also depicted with a ram's head as seen in a relief from Luxor (Pritchard 1958:fig. 162).

The Awassi sheep is usually white, with brown head and feet; black heads and feet, or an entirely white coat, are frequent. However, entirely black or gray, as well as dappled sheep, are uncommon (Hirsch 1933:15), which might explain why in the biblical story Laban agreed so willingly to Jacob's offer: "I will tend your flocks and be in charge of them as before, if you will do what I suggest. Let me pass through all your flock today, removing from it every speckled and spotted sheep and every black lamb, and the spotted and speckled among the goats; and such shall be my wages" (Gen 30:31–32).

The process of domestication led to animals becoming less resistant to diseases. To make matters worse, when they are herded by man, more animals live together in the open country than ever before, a situation that makes flocks even more susceptible (Nissen 1988:25).

Sheep suffer from numerous diseases caused by viruses and other organisms, only a few of which are also common among goats. The most common of these are anthrax, anplasmosis, scabies, strongylosis (liverfluke), and sturdy (Hirsch 1933:23). Other dangerous diseases include brucellosis, contagious agalactia, enzootic virus abortion, foot-and-mouth disease, mastitis, sheep pox, tetanus, and virulent foot-rot (Epstein 1985:57–66). Preindustrial societies could not cope with these diseases and relied on divine intervention and useless practices such as branding. Palestinian Bedouin and Fellahin confirmed that in normal years flock mortality reached 15% to 20% and 40% to 50% in bad years (Hirsch 1933:24). Additional problems plaguing the herds are ticks and lice (Hirsch 1933:23) and a host of flies and worms (Epstein 1985:66–72).

The Biblical references to flock diseases are very few and are in general terms. The curses accompanying the Covenant include the broad statement, "Cursed shall be . . . the increase of your cattle and the issue of your flock" (Deut 28:18), but there is no enumeration as in the curses concerning humans or crops. A somewhat more detailed description appears in the story of the fifth plague, when Moses warns Pharaoh that "the Lord will strike with a deadly pestilence (*deber*) your livestock out in the country, the horses and donkeys, the camels, cattle, and sheep with a devastating pestilence" (Exod 9:3). While the pestilence allegedly distinguished between Egyptian and Israelite herds and did not inflict the Israelites' animals, the maggots or gnats (*kinnîm*) of the third plague (Exod 8:13) and the boils (*šĕḥîn*) of the sixth plague (Exod 9:10) did not distinguish between humans and animals. Thus, the information available does not enable the precise identification of the diseases.

That sheep occupied a central place in the life of the Israelites can be assumed by the different terms employed in the Bible when referring to this animal. The most common term for a sheep is *kebeś* (Exod 29:39; pl. *kĕbāśîm*, Lev 14:10);[49] for a ram, *'ayil* (Gen 15:9);[50] a ewe, *rāḥēl* (Isa 53:7, pl. *rĕḥēlîm*, Song 6:6); a young sheep, *śeh* (Exod 12:5).[51]

The super powers in the Ancient Near East realized the potential of Syria-Palestine and used it as a source not only for finished products but also for the acquisition of livestock. Ancient Egyptian records show that small cattle, which must have included sheep, were part of the tribute and booty resulting from military actions. Thutmoses III (1490–1436 B.C.E.) after the battle at Megiddo, reports that the enemies brought "tribute of . . . large and small cattle for the army of his majesty" (Pritchard 1969a:237), which included "1,929 cows, 2,000 goats, and 20,500 sheep" (Pritchard 1969a:237). Similar results are reported by Amen-hotep II (1447–1421 B.C.E.) after the battles of Aphek (Pritchard 1969a:246) and Anaharath (Pritchard 1969a:247).

The Assyrians did not lag behind. During Iron Age II, they used Palestine as their source of animals. In his eleventh year, in the Bull Inscription, Shalmaneser III (858–824 B.C.E.) states that "At that time I received the tribute . . . silver, gold, tin, wine, large cattle, sheep, garments, linen" (Pritchard 1969a:280).

Somewhat later, Tiglath-Pileser III (744–727 B.C.E.), in the annals of a campaign to Syria and Palestine, states that the booty included "horses, mules, large and small cattle, (male) camels, female camels with their foals" (Pritchard 1969a:283). After his ninth-year campaign against Syria, Arabia, and Palestine, he reports: "I brought away as prisoners 800 (of its) inhabitants with their possessions,"their large (and) small cattle" (Pritchard 1969a:283). A relief from one of this king's palaces shows him taking sheep (and other animals) as booty (Frankfort 1954:pl. 94).

Sennacherib (704–681 B.C.E.), in the report of his campaign in Judah that appears on the Oriental Institute Prism, states that "I drove out (of them) 200,150 people, young and old, male and female, horses, mules, donkeys, camels, big and small cattle beyond counting and considered them booty" (Pritchard 1969a:288). Following him, Esarhadon (680–669 B.C.E.) reports on Prism A about his campaign in Syria-Palestine and says that from Sidon "I drove to Assyria his teeming people which could not be counted, (also) large and small cattle and donkeys" (Pritchard 1969a:290). On Prism B, he records his campaign in Syria-Palestine stating that from Sidon "I led to Assyria his teeming subjects, which could not be counted, (and) large and small cattle and donkeys in great quantities" (Pritchard 1969a:291).[52] That the Assyrians depleted the country of young small cattle has been demonstrated through osteological studies conducted at Tell Jemmeh and Tel Miqne-Eqron (see above).

Sheep, like goats, have been raised for milk, meat, and dung; these products have been described above. Their wool, however, is of greater importance than is the hair of goats.

Wool

The wool of the Awassi sheep is not of the best quality and is used today mostly for coarse fabrics. The average annual wool production of a ewe is 1.75 kg, that of a ram is 2.25 kg, a yearling yields 1.40 kg, and a lamb 0.50 kg (Hirsch 1933:9). While the quality and quantity of wool in biblical times could not have been better, references in the Bible indicate that wool was used for weaving fabrics and making clothes (Lev 13:47; Ezek 34:3; Prov 31:13). Supporting this assertion are large numbers of objects related to weaving (loom weights, spindles and whorls, bone spatulas) found in archaeological excavations that illustrate the wide extent of weaving. However, the objects themselves are not an indicator of how much wool was really involved in weaving, because flax was also a major source of fiber (Borowski 1987:98–9).

Understanding the relationship between the age of livestock to wool production, and its use in weaving can be helped by some finds from Tell Halif in the southern Shephelah. Preliminary examination of bones collected in 1993 from an Iron Age II destruction level of domestic structures inside the fortified city show that 98% of all meat came from caprids (sheep and goats) and cows. While caprids yielded 87% of the identifiable bones, cows represented only 11%. Of the unidentifiable mammal remains, medium-size bones (probably caprids) account for 91%, while large-size bones (probably cows) represent only 8%. The ratio of sheep to goat remains is 2.2:1. Although this might represent a culinary preference, age curves for Iron Age II caprid remains from previous seasons point out that 15% of the caprids were over four years old at the time of death. Survival of caprids beyond two to three years strongly suggests maintenance of sheep for wool (Arter 1995). The discovery of a cache of over 300 loom weights and other weaving-related paraphernalia in 1979 and 1980, and the discovery of another cache of over 100 loom weights in 1992, indicates that wool and textile production were an important part of Tell Halif's economy. It is highly possible that some of this wool and finished products ended up in Assyria as part of Hezekiah's tribute.

Shearing of wool (*gēz*, Deut 18:4) is done once a year in April or May, just before the summer heat commences. If water is available, the sheep are washed beforehand (Song 4:2; 6:6); otherwise, the wool is sold by the fleece (*gizzâh*, Judg 6:37–40) and not by weight, since it is dirty and heavy. Shearing scissors were invented in the Iron Age, around 1000 B.C.E., when iron, which is more flexible than bronze, became more available. Before that, the wool was pulled rather than shorn (Forbes 1956:8). A single man using very simple scissors can shear (*gôzēz*, 1 Sam 25:4) twenty to thirty head per day (Hirsch 1933:29–30). Shearing was an event that brought together many people who were engaged in controlling and shearing the animals, and like during the gathering of other crops, it was an occasion for great celebration (1 Sam 25; 2 Sam 13:23–28) during which food and drinks were offered.

Dyed wool and wool garments were prized by the Assyrians as booty.[53] Tiglath-Pileser III (744–727 B.C.E.), in describing his campaign to Syria and Palestine, includes in the booty lists "garments of their native (industries) (being made of) dark purple wool . . . " (Pritchard 1969a:282). And in his annals from an unknown year, he describes a campaign to Syria and Palestine where he took as booty "blue-dyed wool, purple-dyed wool . . . also lambs whose stretched hides were dyed purple . . . " (Pritchard 1969a:283).[54]

~

The contribution of small cattle to the general economy of the Ancient Near East and to the Israelite economy in particular cannot be overstated. Written records, artistic representations, material culture and osteological studies show how extensively dependent the region was on sheep and goats.

LARGE CATTLE

The discussion of large cattle is divided into two sections; "Cows and Bulls" deals with the animal as a milk and meat producer, "Oxen" with the animal as a beast of burden.

Cows and Bulls

Domesticated cows (*Bos taurus*) evolved from one of two wild species, the now extinct giant ox (aurochs) *Bos primigenius* or the Asian *Bos namadicus*.[55] Domestication took place independently at a number of places, one of which is Çatal Hüyük, in Turkey. At this site, cattle bones dated to 6400 B.C.E. seem to be among the earliest evidence for the domestication of this animal. Another location was Mehrgarh, in the Indus Valley, where domesticated cattle bones from the sixth millennium B.C.E. were discovered (Anthony 1984:48). Evidence of domesticated cattle in early historical times in Mesopotamia is available from artistic representations. Hunting of wild bulls is seen on a gold platter (Fig 2.8) dated to 1450–1365 B.C.E. from Ras Shamra (Aynard 1972:fig 183). An early depiction of cattle as part of a mixed herd, with sheep and goats, can be seen on the "peace panel" of the Ur Standard dated to Ur III, ca. twenty-fifth century B.C.E. (Pritchard 1969b:fig. 304). Here they are seen as part of a procession. Another procession of bulls is known from the temple of Ninhursag in el-Obeid, also dated to the middle of the third millennium. It depicts a row of five longhorn bulls made of shell and black shale (Pritchard 1969b:fig. 98). Another frieze from the same site shows cow milking and milk processing (Pritchard 1969b:fig. 99).

Figure 2.8. *Hunting wild bulls and gazelles. Gold platter, Ras Shamra.* (Musée du Louvre/Antiquités Orientales.)

Recent studies show that cattle domestication in Egypt was quite complex. In Upper Egypt, domestication began only in the Amratian period, and the only type of pre-dynastic breed was the longhorn *B. primigenius*,[56] which often had forward-pointing horns. Since the very beginning of the Old Kingdom, there was a change in the bovine stock, a consequence of the massive imports of Nubian longhorn cattle for fattening. This Nubian breed was not the same as the *B. primigenius* cows used for milk and plowing. A very early illustration of the Egyptian breed is on Narmer's palette (First Dynasty), which depicts in the upper register of both sides the goddess Hathor with long, curved horns, and at the bottom register on the back side a wild, long and curved-horn bull attacking a city and stomping on a human (Pritchard 1969b:figs. 296–7). Longhorn cows were brought to Egypt from different regions, as seen in a relief in the tomb of Ukh-hotep (ca. 1971–1928 B.C.E.), that depicts a skinny desert herdsman leading three oxen as an offering (Pritchard 1969b:fig. 101). However, Egyptian hornless cows and oxen were most likely individuals from the same breed that, because of this trait, had just been culled by their breeders (Muzzolini 1983:55).[57] One such specimen is seen in an Egyptian relief from Deir el-Bahri, dated to 2135–2000 B.C.E., which depicts a man milking a hornless cow directly into a bottle-like vessel. A calf is tied with a rope to the cow's front left leg, and a tear is seen dropping from the cow's right eye (Pritchard 1969b:fig. 100), possibly expressing the artist's understanding of a mother's feelings when she is not able to suckle her offspring.[58] By using sculpture, paintings, and hieroglyphic signs, Zeuner distinguishes four breeds in Egypt, all descended from *B. primigenius*— long-horned, short-horned, lyre-horned, and hornless (in Carrington 1972:71–2).

The existence of long-horned cattle, possibly *B. primigenius*, in Palestine in general and the Araba in particular, can be illustrated by the depiction of a long-horned bull on a cult stand (censor) from the Edomite temple at 'En Hazeva dated to the seventh-sixth century B.C.E. (Cohen and Yisrael 1995:12).

Milk Cattle

Cattle (*bāqār*, Gen 12:16) were initially domesticated for milk, meat, hide, bone, and dung (Fig. 2.9); only later did they become draft animals.[59] Unlike sheep and goats, cattle do not produce fibers for spinning and weaving.

Figure 2.9. *Arab cow.* (Photograph by the author.)

During the Iron Age, cattle were raised primarily for traction and for their milk and dung,[60] and secondarily for meat, hide, and other by-products that become available only after killing the animal.[61] As a draft animal, its meat and other by-products also become available once the animal is judged to be too old for work.

Cows are not given to wandering like sheep and goats, thus their presence in the zooarchaeological record usually indicates a more settled environment, namely that of a village or town.[62] Cattle need more attention than other ruminants and do best under stable conditions. According to Bodenheimer, in preindustrial Palestine there were three native types which probably were very close to those extant in the Iron Age (Bodenheimer 1935:118–22). The most common and adapted breed is the small, furry Arab cow (Fig. 2.10a). It is very lean, with a brown to black or black-white body and poorly developed musculature and udders. It weighs about 260 kg and gives 400 to 700 liters of milk per year [63] during a four- or five-month period when the calf is present. The bulls and oxen may be twice as big as the cow. This animal is a working animal whose meat is of very poor quality. It is adapted to semiarid conditions; like a goat it is easily satisfied and fairly resistant to diseases (Bodenheimer 1935:120).

The second breed is the Beirut cow (Fig. 2.10b), which is actually a better-bred form of the Arab cow. Its yellow-brown to brown body is larger, weighing 230–350 kg. It yields 1,500–2,000 liters per year during a period of seven or eight months when the calf is present. Selected cows may give up to 4,000 liters under favorable conditions. This breed is relatively resistant to diseases, but requires supplementary feed (Bodenheimer 1935:120–21). Its predecessor may have been the cow used in the Red Heifer ritual.

The best milk-producing native cow is the Damascus cow (Fig. 2.10c), which is native to Syria and is identical in appearance to the Egyptian cow goddess Hathor (Pritchard 1969b:fig 573:8). This reddish to dark brown cow

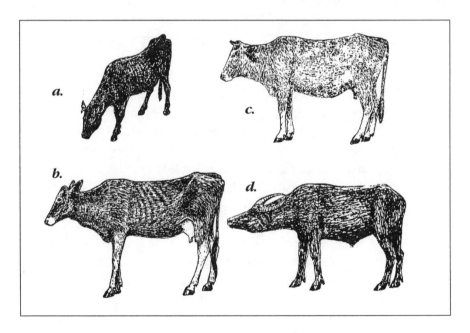

Figure 2.10. *Bovids.* (From Bodenheimer 1935:pl. xii.)
a. Arab cow.; b. Beirut cow.; c. Damascus cow.; d. Buffalo.

has a smooth hide and weighs 380–500 kg. It is raised only where irrigation for growing feed is available. Its yield is 3,000 liter per year during a period of seven or eight months when the calf is present. Good cows may produce as much as 5,000 liters. This breed is not fit for any work, requires good nutrition, and is not resistant to climatic changes and diseases (Bodenheimer 1935:121).[64] Cross-breeding between the Beirut and Damascus cows produced a more resistant milk-yielding cow of 2,500–3,000 liters per year with better quality beef.

Another breed extant in the northern region is the Golan cow, which yields 800–1,000 liters per year, has good beef, is sturdy and resistant, and can be used as a draft animal (Bodenheimer 1935:121).[65]

In general, it is very difficult to determine from zooarchaeological remains whether the animals were originally used for work or for milk. Nevertheless, it seems that at the end they were all destined to be consumed. Cattle skeletal remains were identified at several Iron Age sites, including Ashdod (Haas 1971), Dan (Wapnish 1993:430–5), Halif (Arter 1995), Lachish (Lernau 1975), and Jerusalem (Lernau 1995). The finds at the latter site have been related by the excavator, K. E. Kenyon, to cultic rituals and the reforms of Josiah.[66]

There are several diseases that affect cattle, the most destructive of which are anthrax, pirplasmosis, foot and mouth disease, and cattle plague. Each of these can bring total devastation of the herds and could not be controlled under conditions prevalent during preindustrial times. Only early quarantine could arrest their deadly effects. Biblical terminology does not distinguish between animal diseases, and they are all referred to as *deber* (pestilence) common to humans and animals alike (Exod 9:3; Jer 21:6; Ezek 14:19, 21).

As depicted on Mesopotamian cylinder seals (Aynard 1972:52 and fig. 26) and bas-reliefs, cattle were kept in sheds (*rĕpātîm*, Hab 3:17), where milking and milk processing took place. One of the dairy scenes is a bas-relief from the temple of Al 'Ubaid (beginning of third millennium B.C.E.) that shows in the center a reed shed "from which emerge the animals' forequarters (Fig. 2.11). To the right, men seated on low stools placed behind the animals are milking two cows. A young calf licks its mother" (Aynard 1972:52). The scene shows one man swinging a large jar (or possibly a goatskin) back and forth to churn the milk,[67] which is then separated by filtering, while another man fills a jar with butter.[68]

Figure 2.11. *Panel showing men milking cows. Painted plaster panel from the Temple of al 'Ubaid, Iraq.* (Collections of the Henry Ford Museum and Greenfield Village, Dearborn, Michigan.)

Cattle in Canaan and the Ancient Near East

Large cattle, together with small cattle, he- and she-asses, servants, camels, and tents, were always considered a measure of wealth (Gen 12:16; 13:5).[69] Large herds of cattle needed large tracts of land for grazing as alluded to in the stories about the conflict between Abraham's and Lot's herds (Gen 13) and about Jacob's family in the Land of Goshen (Gen 45:10; 47:60). When the Israelite monarchy was established, the king became owner of large herds of cattle. According to the biblical account, under David special overseers were appointed for the *bāqār*, "Shitrai from Sharon was in charge of the herds grazing in Sharon, Shaphat son of Adlai of the herds in the valleys" (1 Chr 27:29). The cattle were probably used as track animals, and for their by-products, especially milk, meat, and hide.[70] It is possible that dung from the royal animal sheds was used in fertilizing the royal fields, orchards, and gardens.

In general, young calves were considered a culinary delicacy and were prepared for special meals. According to the story, when the three guests came to Abraham to announce the imminent birth of Isaac, he offered them a meal including a young calf (*bēn bāqār*), butter, fresh milk, and baked goods

(Gen 18:6–8).[71] Calves, specially selected for fattening (*ʿēgel marbēq*), were served on particular occasions, such as when Saul visited the spiritualist woman at En-Dor (1 Sam 28:24). Young calves were also considered choice animals for sacrifice (Lev 9:2; 16:3; Num 7).[72]

The bull (*pār*) was the symbol of power and fertility. It is mentioned in the Hebrew Scriptures numerous times, the majority of which are in connection with sacrifices. Sometimes the Bible uses the bull as a symbol for YHWH and in this context employs the terms *ʾabîr yaʿăkob* (Gen 49:24; Isa 49:26) or *ʾabîr yiśrāʾēl* (Isa 1:24). In the Ancient Near East, the bull was the symbol of various gods in several cultures (Pritchard 1969b:figs 828 [Ras Shamra], 832 [Hazor]),[73] a fact which must have influenced Israelite iconography and was expressed verbally and iconographically. The story of the Golden Calf (Exod 32) is one expression of this influence. In reality, the bronze bull statuette from the Bull Site in the Samaria hill-country is a remnant of such cultic use of the bull in early Israelite religion. Architectural remains uncovered at this site included a large stone circle with a *maṣṣēbâh* (standing stone) on its eastern side; the statuette was found there. Mazar, the excavator, suggests that "the figurine was probably used by Israelite settlers in this region of the northern Samarian hills" (1990:352).

Female cattle are mentioned significantly fewer times, and mostly metaphorically. In Genesis 41, cows appear in Pharaoh's dream symbolizing good and bad agricultural years. In Hosea 4:16, Israel is compared to a stubborn cow, and in Amos 4:1, northern Israelite women are ridiculed when referred to as "Bashan cows." Cows were harnessed for work. The biblical account mentions that cows were used for pulling the wagon bringing the ark from Philistia back to Judah (1 Sam 6) and that they were sacrificed when they reached their destination. When the red cow was selected for the "Red Heifer ritual," she was supposed to be "without blemish or defect, one which has never borne a yoke" (Num 19:2).[74]

The motherly emotions of cows mentioned above as depicted in an Egyptian relief are also illustrated in other ways. The close relationship between the nursing mother and her calf is depicted in the famous ivory carvings from Arslan Tash (Fig. 2.12) that shows a cow nudging or licking the root of the tail of a suckling calf (Frankfort 1954:fig. 169b; Davidson 1962:204). This recurring motif depicts an actual mode of behavior

Figure 2.12. *Cow suckling a calf. Ivory inlay from Arslan Tash.*
(The Bible Lands Museum, Jerusalem.)

characteristic of nursing cows and other ruminants. By doing so, the mother encourages the young animal to suckle. Most of the ivories of this type are executed with great talent and show the loving mother being concerned and caring for her young. A line drawing on a pithos from Kuntilat 'Ajrud (second half of the ninth century B.C.E.) attempts to depict the same scene, but in a different medium (Meshel 1978; Meshel 1993:1462). Such depictions could have originated only in a cultural environment steeped in herding where intimate knowledge of this behavior existed.

Syria-Palestine was always animal rich, and the Egyptians were well aware of the animal resources available there. This was exploited by the several pharaohs of the New Kingdom, such as Thutmoses III (1490–1436 B.C.E.), who after the battle at Megiddo received tribute of "large and small cattle for the army of his majesty . . . " which included "1,929 cows, 2,000 goats, and 20,500 sheep" (Pritchard 1969a:237). His successor, Amen-hotep II (1447–1421 B.C.E.) was not as successful because, on returning from his Asiatic campaign, he reported on the Memphis and Karnak Stelae that from Shamash-Edom at Retenu (Canaan) he took as booty twenty-two cattle[75] (Pritchard

1969a:245). From his battle at Aphek in Year 9, he brought booty that included cattle, and near Mt. Hermon he captured and killed a few people and had sixty cattle be driven before his chariot (Pritchard 1969a:246). However, on his return to Memphis the list of booty does not mention the cattle (Pritchard 1969a:246). Possibly it was slaughtered on the march and fed to the troops. His later campaigns were more successful as far as cattle acquisition was concerned. From his battle at Anaharath, he took booty including "bulls: 443; cows: 370; and all (kinds of) cattle without limit," which he reported on his return to Memphis as including "all (kinds of) cattle, without their limit" (Pritchard 1969a:247). But not only live animals were either brought or traded from Canaan. The story of Wen-Amon (ca. 1100 B.C.E.) records that among the goods sent to Egypt were "500 cowhides" (Pritchard 1969a:28).

On this score, in their turn the Assyrians did not fall behind. As with other animals, the Assyrians used their military expeditions and military victories in Syria-Palestine to replenish their herds of cattle. Ashurnarsipal II's (883–859 B.C.E.) expedition to Carchemish and the Lebanon brought back as tribute "1,000 (heads of big) cattle" (Pritchard 1969a:276) and Shalmaneser III (858–824 B.C.E.) reports in the annals of his first year on the so-called Monolith Inscriptions that "I received the tribute . . . silver, gold, large and small cattle, wine" (Pritchard 1969a:277). He repeats this boast on several other occasions, and in his eleventh year Bull Inscription he enumerates tribute he received and includes large cattle (Pritchard 1969a:280).

Tiglath-Pileser III (744–727 B.C.E.) reports in the annals from an unknown year that his booty from a campaign to Syria and Palestine included livestock of large cattle. Then, from his ninth year campaign against Syria, Arabia, and Palestine he also brought large cattle (Pritchard 1969a:283). The enormous size of the herds in that period is exemplified by the next report. In one of his campaigns in an unknown year in Syria, Palestine, and Arabia, he defeated Samsi, queen of Arabia and killed, among other creatures, 20,000 head of cattle (Pritchard 1969a:284).

Similar reports are available from the succeeding kings. Sennacherib (704–681 B.C.E.) brought back from his campaign in Judah "big and small cattle beyond counting and considered them booty" (Pritchard 1969a:288). Esarhadon (680–669 B.C.E.) on Prism B from his campaign in Syria-Palestine, states that from Sidon "I led to Assyria . . . large and small cattle and donkeys in great quantities" (Pritchard 1969a:291).[76] Ashurbanipal (688–633 B.C.E.) followed in his footsteps and from the campaign against Arabia he brought back "large and small cattle" (Pritchard 1969a:299). In the contest for who exploited more of the faunal resources of Syria-Palestine, the Assyrians won.

The fact that Egyptian and Assyrian armies continuously used Syria-Palestine as a source for acquiring herd animals strongly suggests on one hand that animal breeding was a very successful enterprise that could hardly be depleted, but on the other that animal confiscation and deportation by political entities in power was done with some kind of plan with an eye for future campaigns.

Notes

1. See Borowski 1987:30.
2. Since Gideon was engaged in threshing wheat, which is a late spring–early summer activity, the kid might have not been weaned; see "Small Cattle: Meat" below.
3. This is strongly supported by zooarchaeological studies showing that the number and percentage of caprid (sheep and goats) bones is very high, Str. III–51.7%; Str. II–46.1%; Str. I–67.4% The high percentage of large cattle indicates that the community living in this settlement was relatively settled. The large number of adult animals shows that the emphasis at the site was on exploitation for milk, wool, and labor, and not just for meat (Hellwing and Adjeman 1986).
4. A similar preference was also observed at the coastal site of Tel Michal (Hellwing and Feig 1988:246).
5. A situation very similar to that of Bethuel, Rebecca's father (Gen 24).
6. The Bedouin term for 'tent' is *bit ša'ar* 'house of hair,' and there is a possibility that *bayit*, the Hebrew reference used here actually means 'home' and could refer to 'tent.'
7. These numbers might appear enormous, but compared with the 347,394 sheep listed on an Ur III tablet, they are reasonable (Postgate 1986:198 n. 13). In Ebla, six-figure numbers of sheep in herds are also quite common (von Soden 1994:94–5).
8. An example of a mixed herd is the one Jacob sent as a gift to his brother, Esau: "two hundred she-goats, twenty he-goats, two hundred ewes and twenty rams, thirty milch-camels with their young, forty cows and ten young bulls, twenty she-donkeys and ten donkeys" (Gen 32:14–15).
9. The last large wandering resulting from drought occurred in 1947 and brought the Negev herds to the north even further than the Jezreel and Beisan valleys. After 1948, there was no repetition of such movement because of the political situation.
10. Settling in Sodom, and becoming an urbanite, might have forced Lot to start tending herds under the sedentary system.

11. Interestingly, Mazar reports that during the period of the Israelite Settlement in Giloh, the bone collection (which was meager) included six cow bones, two belonging to caprids, and two to donkeys (1990:90). Although the sample might be skewed, having large cattle at this site shows a certain degree of permanence even during this period.

12. By doing so, the shepherd proved to the herd owner that he was not negligent. A famous rescue tale is that of the shepherd who in the attempt to save his goat found the Dead Sea Scrolls.

13. In Amos 7:14, the prophet refers to himself as *bôqēr* 'cattle-herder.'

14. The shepherd was supposed to try and save the sheep from a mauling animal, or at least rescue "a pair of shin-bones or the tip of an ear" (Amos 3:12) to prove to the owner that he indeed did his job.

15. See below, "Equids: Mule".

16. Compare this to Judges 5:25: "He [Sisera] asked for water: she [Jael] gave him milk . . . ."

17. Sheep come into heat between June and September; their period of gestation is 152 days and lambing starts in December, lasting until April. The height is December through February. Goats give birth from December to April, mainly in January and February. As a rule, at an animal's first birth, does give only one kid, but afterward twins account for half the births in a herd (Hirsch, 1933:59).

18. Depending on grazing conditions, optimal weaning is done after two or three months, in March or April. At this point, the average weight of male lambs is 20 kg. Weaning at the age of two to three months produces maximum meat with minimum investment and allows a milking period of three to four, and sometimes up to five, months.

19. From the root RBQ, 'to tie fast' (Brown et al. 1906:918); see also the discussion on fattened animals.

20. This dry cheese, which can be preserved for several years (Pravolotsky and Pravolotsky 1979:65), may be ground or kept in lumps and mixed with water when needed. This type of cheese is sometimes refered to by the Arabic term *kishik* (Hirsch 1933:39; Avitsur 1972:230–1).

21. For traditional vs. improved methods, see Ishak 1987.

22. While goatskins must have been used for churning throughout history, they did not survive in archaeological context. The same is true for gourds. However, a particular vessel-type from the Chalcolithic period is identified as a "butter churn" (Amiran, 1963:52–3). Similar vessels from the Hellenistic period were found at Shiqmona (Avitsur, 1976:fig. 181). Other clay vessels must have been used along with goatskins during various periods.

23. The Bedouin of Sinai tend to produce *lebben* by churning (Pravolotsky and Pravolotsky 1979:64). Hirsch and Avitsur suggest that milk for *lebben* is first boiled before processing (Avitsur, 1972:230; Hirsch, 1933:39–40). According to Krispill, *lebben* is the leftover water after making zibdeh (Krispill 1986:29).

24. Possibly *mîṣ ḥālāb* (Prov 30:33).
25. Sometimes cheese is made from buffalo milk (Ishak 1987:221).
26. An extract from the stomach of a ruminant used to curdle milk.
27. There is no evidence in the Bible or otherwise that the Israelites smoked, dried, or salted meat.
28. This is used for culling the herd, which can also be achieved by selling the young animals.
29. The biblical references to a "male lamb a year old" are too numerous to cite.
30. See also the story about Gideon offering a meal to the divine messenger at the time of the wheat harvest (Judg 6).
31. Other references to the prescription are found in Lev 9:3; 12:6; 14:10; 23:19; Num 28:11, 19, 27; 29:2, 8, 36.
32. From the root *MḤḤ*, "be fat, contain marrow," Akk. *muḫḫu* (Brown, et al., 1906:562; Koehler & Baumgartner, 1953:512).
33. From *kar* 'lush pasture' or *KRR* 'round enclosure' (Brown, et al. 1906:499).
34. Other terms for fattened animals are *ʾēyl milluʾîm* (fattened ram, Exod 29:26, 31–2), *ʿēgel marbēq* (fattened calf, 1 Sam 28:24), *měrîʾ* (fattened ox, 2 Sam 6:13), *barburîm ʾabûsîm* (fattened birds, 1 Kgs 5:3).
35. Meat was never eaten raw (*nāʾ*, Exod 12:9) and there was a strict prohibition against blood consumption (Gen 9:4). This prohibition was well known; see the incident mentioned in 1 Sam 14:32–4. Meat of a dead (*něbēlâh*, Lev 7:24) or a devoured animal (*ṭěrēpâh*, Exod 22:30) was not allowed to be consumed by humans, but could be fed to dogs. Meat not fit for consumption was known as *piggûl* (Lev 7:18).
36. There are several terms for cooking and boiling utensils (Num 11:8; 2 Kgs 4:38; 1 Sam 2:14; Micah 3:3). Other cooking installations and utensils are mentioned in Lev 7:9; 1 Sam 2:13.
37. Many reasons have been offered to explain this prohibition, but none is decisive.
38. For the use of dung in agriculture, see Borowski 1987:145–6.
39. In Babylonia, dried cow dung was kept in piles to be used as fuel (von Soden 1994:94).
40. For example, 96% of the bones uncovered at Mount Ebal belonged to ruminants, and none of the bones were worked (Hess 1993:136). Most of the bones found at the Ophel (84% in Building D and 85% in Building C) belong to sheep and goats and a small number to cattle. A very small number of the bones had been worked (Kolska-Horwitz and Tchernov 1989).
41. Although a late example, one such instrument was discovered at a 70 C.E. destruction layer in Jerusalem (Ariel 1990:142–3). Another site where worked bones from different periods were discovered is the coastal city of Ashkelon (Wapnish, 1991).

42. In zooarchaeological studies, this figure is referred to as "minimum number of individuals" (m.n.i.).

43. Sumerian cylinder seals show goats with an appearance similar to that of the modern goat (long ears and long hair), see also Hirsch 1933:55.

44. Under laboratory conditions, Bedouin goats lasted for two weeks without water and did not lose their appetite (Shkolnik, 1977:102).

45. One term referring to goats is *bĕʿîr* (Exod 22:4; Ps 78:48), from the root *BʿR* 'to burn,' possibly relating the animal to the damage it causes.

46. Goat hair was used, at least since Roman times, for making coarse cloths, cloaks, rugs and felt slippers (Forbes 1956:58).

47. Something similar is seen in one of Sennacherib's reliefs (Yadin 1963:302).

48. For wild sheep, see Nissen 1988:fig 6.

49. Sometimes it appear as *keśeb* (Lev 3:7; pl. *kĕśābîm*, Gen 30:32; Lev 1:10) or *kibśâh* (2 Sam 12:3; pl. *kibśôt*, Gen 21:28 or *kĕbāśôt*, Gen 21:30).

50. At times, *ʿattûd*, which is a he-goat, is used also for ram (Gen 31:10).

51. The term *śeh* is used sometimes to designate any young small or large cattle, as in Leviticus 22:28 and Deuteronomy14:4.

52. Ashurbanipal (688–633 B.C.E.) had a similar experience in Arabia (Pritchard, 1969a:299–300).

53. The Assyrians were attracted to woolen and other garments (Elat 1977:83–97) and Hezekiah's tribute to Sennacherib was partially paid with wool (Elat 1977:219).

54. For a discussion on dyeing, see "Molluscs: Murex" below.

55. Bodenheimer suggests that the aurochs, which is the ancestor of all Palestinian cattle as well as of the zebus, is the Assyrian *rimu* and biblical *rĕʾēm* that was hunted by the Assyrian kings (Bodenheimer 1960:51, 103). *B. primigenius* is depicted by Mesopotamian artists from earliest times onward (Aynard 1972:45). Other wild forms of cattle occurring in ancient Mesopotamia were the arni-buffalo (*Bubalus bubalis*) and the wisent (*Bison bonasus*) (Bodenheimer 1960:102).

56. Remains of *B. primigenius* from the Pleistocene were also found in caves in southern Syria and northern Palestine (Bodenheimer 1935:36–7; Bodenheimer 1960:22).

57. Short-horned *B. brachyceros* did not appear before the New Kingdom, nor did the short-horned zebus, imported from Asia (Muzzolini 1983:55). Clutton-Brock reviews the complexity of cattle domestication in Egypt and North Africa and also suggests that the earliest appearance of humped cattle in Egypt is dated to ca. 1400 B.P. (Clutton-Brock 1989:204).

58. This motherly emotion, which is actually human, should be compared with well-known Palestinian depictions of mother and calf (see below).

59. For cattle as draft animals, see "Other Draft and Pack Animals" below.

60. For the use of dung, see Borowski 1987:145–6.

61. It is quite possible that the "200 [leather] coats of mail belonging to his [Prince of Megiddo] wretched army" (Pritchard 1969a:237) taken as booty by Thutmoses III (ca. 1468 B.C.E.) were made of cowhides or hides of similar animals.

62. In Iron Age I Ashdod (Str. XII), one of the Philistines' urban centers, the faunal inventory suggests a ratio of cattle to sheep and goats in favor of cattle (Kolska-Horwitz 1993). This is probably a reflection of the urban mode of life.

63. Avitsur claims that the Arab cow produces 500–600 liters milk per year (Avitsur 1972:230).

64. The reference to "the cows of Bashan" in Amos 4:1 is probably to this breed because they are big, fat, and do not work.

65. Preindustrial Palestine also had domestic buffaloes (*Bubalus bubalus*), which will not be discussed here. For details see Bodenheimer 1960:122.

66. On the possible cultic connection, see "Animals in the Cult of Ancient Israel" below.

67. In Canaan, so-called butter churns made of clay were used in the Chalcolithic period.

68. On dairy products see above.

69. See also Gen 20:14; 24:35; 26:14; Deut 8:13. Cattle were considered more valuable than small cattle. One Old Babylonian letter reports that one head of cattle was worth as much as thirty sheep (von Soden 1994:94).

70. Cowhides were goods traded in Iron Age I international commerce, as mentioned in the Wen-Amon story (Pritchard 1969a:28).

71. Abraham was apparently not concerned about mixing meat and milk.

72. For a discussion on fattening animals see above "Small Cattle: Meat."

73. Examples for this are too many to mention, therefore from Canaan I would like to cite here only the one from Ashkelon which is relatively close in time and space to the Israelite period (Stager 1991:3, 6–7). In Egypt, the bull represented the god Apis (Pritchard 1969b:570). It should be mentioned that the cow was also used in Near Eastern iconography to depict goddesses, e.g. the Egyptian Hathor. Other cultures also revered the bull. Some examples of earlier depictions and from neighboring cultures are in order. See Davidson for bulls in Crete and in Minoan art (1962:267, 268–9) and for calves in Crete on sarcophagus paintings from Hagia Triada (1962:260–1).

74. The "Red Heifer ritual" was performed for purification. The selected red cow might have been a forerunner of the Beirut cow because of its color and its potential to be used for work.

75. In Karnak, it says 18+19.

76. On Prism A he describes a campaign in Syria-Palestine in which from Sidon "I drove to Assyria . . . large and small cattle and donkeys" (Pritchard 1969a:290).

Chapter 3

Draft and Pack Animals

EQUIDS

The Equidae family (Fig. 3.1) is a four-legged, ungulate (hoofed) mammalian group, which, according to Groves, is divided into two subgroups—horses and zebras, and donkeys and onagers (an Asian wild ass) (1986:16). These species can interbreed, but they produce infertile offspring.[1] While the horse and donkey were domesticated and used for riding and as pack and draft animals, the zebra and the onager were never tamed.[2]

Around 8500 B.C.E., at the beginning of the Neolithic period, people changed their mode of life and began to cultivate cereals and domesticate goats, sheep, and, somewhat later, cattle and pigs. This, in part, could be attributed to the increasing aridity of the climate in the Near East. Climatic changes created growing pressure on the food supplies available to the expanding human populations, thus providing a stimulus for developing new ways of obtaining food through domestication. The absence of equid remains in the fossil record suggests that when domestication of livestock first took place there were no wild horses (*Equus ferus*) in the Near East. Although there were wild asses (*Equus africanus* and *Equus hemionus*) in the region,[3] it is possible that they were harder to handle as herd animals than goats, sheep, or cattle. Because of that, equids were not included in the first wave of domestication, and never became a primary source of meat. Remains of the earliest domestic horse were found at archaeological sites dated from around 4000 B.C.E. in the Ukraine, with some scattered remains recorded at early sites in central Germany.[4] The early appearance of domestic donkeys can be

Figure 3.1. *Equids*. (Anthony 1984:15.)

seen on wall paintings and in burials from Egypt and the Near East dated from third millennium B.C.E. (Clutton-Brock 1992:11).

In Canaan, the possible appearance and use of pack animals during the fourth and third millennia B.C.E. can be connected to population expansion into the Negev and Sinai, which was related to increasing trade in general, the formation of trade links with pre-dynastic Egypt, and metallurgical activity in that region. Horses became widespread in the Near East, however, during the first half of the second millennium B.C.E., when the earlier forms of solid-wheeled carts were refined, by way of the cross-bar and spoked wheels, into the horse-drawn chariot (Sherratt 1983). The horse was mentioned occasionally in cuneiform records from around 2050 B.C.E. One three-year account tablet records 37 horses (*anše-zi-zi*), 360 onagers, 727 hybrids, and 2,204 donkeys (in Clutton-Brock 1992:90). As time went on, the onager disappeared completely from the written records, while the horse remained an animal of high status and was mentioned more often in the texts. After its appearance, the trainable horse was considered a better animal, and the reason for keeping the untrainable onager diminished. With the diminished use of the onager, the term previously used for the donkey x onager hybrid—*mule*—was redeployed for the new donkey x horse hybrid, the mule (Clutton-Brock 1992:90).[5]

Harnessing and Wheeled Transportation

Pictographs on clay tablets from Uruk in southern Mesopotamia dated to 3200–3100 B.C.E. depict roofed sleds and wheeled carts (Anthony 1984:20). During this period, carts (biblical *ʿăgālâh*, Gen 45:19)[6] were drawn by oxen, which were also used as draft animals during the Early Dynastic period in Ur (ca. 2500 B.C.E.). However, the battle scene on the Ur Standard (2600–2500 B.C.E.) shows wooden four-wheeled wagons drawn by teams of four donkeys or other equids.[7] While early wagons had solid wooden wheels, the earliest spoked wheels (biblical *ʾôpān*, 1 Kgs 7:33) are dated to around 1900 B.C.E. (Clutton-Brock 1992:70). Light single-axled chariots with two spoked wheels (biblical *rekeb*, Judg 4:3; *merkābâh*, 1 Kgs 10:29) became common in Egypt and Western Asia in the first half of the second millennium B.C.E. (Anthony 1984:20), they are well-depicted in reliefs, paintings, and models, and exist as remains of the actual objects.

The common way carts, wagons, and chariots were pulled was by draft animals in pairs or in teams of four harnessed by a yoke (*ʿol,* 1 Sam 6:7)[8] to a harnessing pole. For an equine animal, the yoke harness is less effective than for bovines because of the absence of a suitable point to pull against on the animal's neck or back. The throat-and-girth harness was thus invented to adapt the yoke harness to equine animals. Although the throat-and-girth harness is extremely ancient and is still being used today, it was still a poor adaptation for the horse. Equids differ anatomically from oxen, and their maximum draft power comes from the shoulders and chest rather than the neck, the best horse harness found in antiquity utilized the animal's breast to achieve draft power (Bulliet 1975:178–79, 181).

In earlier times, control of draft animals (including equids) was gained by a nose ring (*ḥāḥ,* 2 Kgs 19:28; Isa 37:29),[9] but later other means were developed including reins (*mośrôt,* Jer 5:5; 27:2; or *ʿăbôt,* Isa 5:18), bridle (*meteg,* Ps 32:9), and bit (*resen,* Isa 30:28; Ps 32:9) (Clutton-Brock 1992:72).[10]

At the beginning of the second millennium B.C.E., throughout Europe and Western Asia, the horse was slowly replacing the ox, not as a common draft animal for wagons and carts but as the power for pulling chariots used in hunting and war (Clutton-Brock 1992:58). That use also gained popularity in Egypt and Assyria and can be seen on numerous artistic representations. The role of the horse in the Ancient Near East in general, and Eretz Yisrael during the Iron Age in particular, was limited to pulling chariots and for riding, mostly in military contexts.

Donkeys

The wild ass was as well known in the Bible as the domesticated donkey (*ḥămôr*) (Fig. 3.2). Job 39:5–8 describes the wild and untamed animal quite accurately: "Who has let the Syrian wild ass (*pereʾ*) go free and loosed the bonds of the Arabian wild ass (*ʿārôd*)? I have made the wilderness his home and the salt land for its dwelling place. He scorns the uproar of the city, he does not obey a driver's shout. He roams the mountains for pasture, and seeks out every patch of green."[11]

While different races of donkeys (*Equus asinus*) have evolved over the last 4,000 years by natural selection in response to local conditions of climate and nutrition,[12] it should be remembered that only one species of the wild ass (*Equus africanus*), which was once widespread as a number of separate subspecies over the whole of Saharan Africa and most likely in Arabia, has been

Figure 3.2. *Donkey* (From Bodenheimer 1935:pl. xiii.)

the progenitor of the domestic donkey (Clutton-Brock 1992:62–3).[13] Its relative, the Asiatic wild ass (*E. hemionus*), especially its Syrian onager subspecies (biblical *pere³*),[14] was hunted in Mesopotamia for sport and its hide. Captured onagers[15] were kept for breeding with domestic donkeys (Postgate 1986:199–200) and horses to produce mules (Clutton-Brock 1992:36–7), but were never domesticated and should not be considered the ancestors of the domestic donkey (*E. asinus*).[16] Equid remains from the "royal" tombs of Ur, Kish, and other sites have been analyzed as belonging to hybrid animals, *E. asinus*, and wild asses—*E. hemionus* as well as *E. caballus* (Zarins 1986). Interpretation of textual evidence from Mesopotamia suggests that, as time went on, the onager disappeared completely as a kept animal, but its appearance in reliefs from Nineveh (645 b.c.e.) shows that hunting onagers as sport, like hunting lions, continued through the reign of Ashurbanipal (Frankfort 1954:fig. 112; Pritchard 1969b:fig. 186).

Of all domestic animals, the African ass is second only to the camel in performance under desert conditions—its ability to graze on desert scrub (Gen 36:24) and its power to go for long distances without drinking (Lydekker 1912:220–1; Dent 1972:31).[17] These characteristics are suggested by a wall painting in a tomb at Beni Hasan in Egypt dated to about 1890 b.c.e. (fig. 3.3), which depicts a group of Asiatics who crossed the Sinai Desert into Egypt leading, among other animals, two donkeys carrying equipment, people, and

Figure 3.3. *Asiatics coming down to Egypt. Wall painting in Tomb 3 at Beni Hasan.* (From Newberry 1893:pl. XXXI).

other burden. The donkeys' ability to withstand the harsh arid conditions along the road made crossing the desert possible.

The popularity of the donkey is demonstrated by the fact that, in preindustrial Syria, there were four distinct breeds—"a light and graceful type with a pleasant, easy gait, used by ladies of rank; a so-called Arab breed, reserved entirely for the saddle, and carefully groomed and tended; a stouter and more clumsily made strain employed for ploughing and other agricultural operations; and, lastly, the large Damascus breed, characterised by its length of body and inordinately long ears" (Lydekker 1912:222). The latter was white, similar to a breed reared in Baghdad, where it was favored because of its color and high speed.

Although the original reason for domesticating the wild ass might have been the consumption of its flesh and milk, it is most likely that the donkey became a favorite because of its ability to carry heavy loads before the invention of the wheel and the introduction of the wagon (Dent 1972:33–4). While the domestic donkey was slower than the horse, because it descended from the wild ass of the hot deserts of Africa and Arabia, it was immune to the harsh climate of the Near East (Fig. 3.4). Moreover, its most important role was to sire the mule (Clutton-Brock 1992:14). It is clear from Sumerian texts that the domestic donkey was used as the common agricultural draft animal, while a small number of onagers were retained for cross-breeding with donkeys to produce large powerful animals that were yoked and used to draw chariots. These hybrid offspring were as large as the early domestic horses (Clutton-Brock 1992:89).

Archaeological and textual evidence, as well as ancient art, indicates that the wild ass was first bred in captivity in Egypt and in Western Asia. Davis has identified ass bones and teeth from the Chalcolithic period (3660–2400 B.C.E.) at Teleilat Ghassul in Jordan and from the Early Bronze Age at Arad in the northern Negev of Israel (Davis 1987:165). During the Old Kingdom (Dent 1972:28), or at the latest by 2500 B.C.E.,[18] the wild ass was in general use, presumably as a beast of burden. At Tal-e Malyan, remains of domestic donkey dated to 2800 B.C.E. were identified, after which date the number of remains in archaeological sites increases. Donkey appearance in Sumerian records, as well as osteological evidence, also increases and indicates that donkeys replaced oxen as draft animals with four and two-wheeled carts.

Figure 3.4. *Camel and donkey with riding saddles.* (Photograph by the author.)

This can especially be seen on the famous "Standard of Ur," where equids, possibly donkeys, are seen used as draft animals.[19] It is also illustrated by an Early Dynastic copper model from Tell ʿAgrab in Mesopotamia, which shows four donkeys pulling a two-wheeled cart (Frankfort 1954:27 and fig. 20a; Pritchard 1969b:fig. 166). Clay figurines from Chalcolithic, Early Bronze Age I, and Early Bronze Age III contexts showing donkeys carrying different types of load reinforce the theory that, by the third and early second millennium B.C.E. the donkey was being harnessed in the Near East (Davis 1987:165). Moreover, donkeys as animals of burden carrying large loads of wheat appear in Egyptian reliefs dated to Dynasties V and VI (Michalowski 1968:figs. 241, 260).

During the second millennium, as horses replaced donkeys and their hybrids as draft animals for the light, spoked-wheel chariot all over Western Asia and Egypt, donkeys and mules continued to serve in agricultural tasks and as pack animals (Clutton-Brock 1992:85).

One piece of evidence that strengthens this point comes from the Late Bronze Age levels at Tell Halif, in the northeastern Negev on the border with the hill country. The site is situated on one branch of the Via Maris, which led from Egypt northward. Large quantities of bones of young donkeys un-covered there strongly suggest that the inhabitants of the site, which served during that period as a caravan stop, were engaged in raising donkeys for the caravan trade (Zeder in Seger et al. 1990:27). The mature donkeys were obtained by the caravans, while the young donkeys whose bones were discovered were culled to maintain herd size and quality. Tell Halif's role as a trade post is underlined by the sharp rise in fish remains, especially of exotic fish species (Zeder in Seger et al. 1990:27).

That donkeys continued to be used as pack animals in this region dur-ing the Iron Age can also be attested by the mention of ṣmd ḥmrm (a pair of pack donkeys) in Arad Letter 2 (Pardee 1978:299–300), and the recovery of donkey bones in Beer-Sheba (Hellwing 1984), Ashdod (Haas 1971), and Tell el-Hesi (Bennett and Schwartz 1989). In Ashkelon, articulated donkey re-mains were uncovered in an Iron Age II stratum (personal observation). This was a port city and a center for commerce, which explains the presence of donkeys at the site used as pack animals for the transport of goods.

Donkey Riding and Packing

Although horse riding became preferred because of its greater speed, donkey and mule riding did not cease completely. Donkey riding influenced the early style of horseback riding, and pictorial evidence from the third millennium shows that riding of both horses and donkeys was done with the rider sitting well-back in the "donkey seat." Because of its anatomy, namely its low with-ers and low carriage of the head and neck, the only way to ride a donkey for any length of time is to sit well-back on the loins. Even then, however, the rider will be jolted by the shock of the moving legs if going at any speed. The donkey is thus quite unsuitable for riding at speed and this would have pre-cluded its use as a mount in hunting or battle (Clutton-Brock 1992:66, espe-cially fig. 4.13). A good illustration of donkey riding is depicted in an early second-millennium-B.C.E. figurine from Syria, now in the Bible Lands Mu-seum in Jerusalem (Fig. 3.5). It shows the rider sitting on the back part of the animal, just above the tail, gripping the flanks with the calves (Merhav 1987:101).[20]

In spite of the horse's popularity in Eastern Asia and Europe, during the Iron Age the domestic donkey remained the common means of transport throughout Egypt and Western Asia.

Figure 3.5. *Donkey and rider.* (The Bible Lands Museum, Jerusalem.)

The Role of the Donkey in Ancient Israel

In Israelite society, donkeys (and oxen) seem to have been the most common, and important, animals since they, and not horses or camels, are the ones enumerated among the prized possessions not to be coveted: "Do not lust after your neighbour's wife; do not covet your neighbour's household, his land, his slave, his slave-girl, his ox, his donkey, or anything that belongs to him" (Deut 5:21). Having a donkey was considered basic (Deut 22:3), thus it was important to take care of it, feed it well (Isa 30:24), and let it rest along with all other employees and animals (Deut 5:14). The importance of this animal is also demonstrated by the fact that, as part of the rules for the renewal of the covenant, an instruction is given to redeem its firstling with a lamb (Exod 34:20). A donkey must have been more valuable than a lamb.

The donkey provided transportation for people (Gen 22:3; Exod 4:20) and was used as a pack animal (Gen 42:27; 44). This is clearly illustrated in Egyptian as well as Mesopotamian art. In a wall painting in a tomb at Beni Hasan in Egypt (19th century B.C.E.), white donkeys are depicted being led by Semites migrating into Egypt. One donkey is laden with smelting equipment and the other carries two children (Pritchard 1969b:249, fig. 3). On a relief from Nineveh, Ashurbanipal (668–633 B.C.E.) depicts prisoners taken from a captured Egyptian city among whom are two children riding on a donkey covered with a blanket-saddle (Pritchard 1969b:fig. 10). While it was employed in basic agricultural tasks such as plowing (Deut 22:10) and threshing (Pritchard 1969b:fig. 89), the donkey was also used as a riding animal by people of high standing, women (Josh 15:18 = Judg 1:14; 1 Sam 25:20) as well as men (2 Sam 17:23; 1 Kgs 2:40), even royalty (2 Sam 19:27). According to the biblical account, David had a special overseer, Jehdeiah the Meronothite, in charge of the donkeys (1 Chr 27:30). Owning young donkeys (*'ăyārîm*) in large numbers was apparently also considered a status symbol (Judg 10:4).[21] However, it seems that the female (*'ātôn*) of the white variety was preferred for riding (Fig. 3.6), as alluded to in Judges 5:10. Other instances where riding she-asses is specifically mentioned are those describing Balaam the Seer (Num 22) and the Shunamite woman (2 Kgs 4:24).[22] Riding a donkey took some preparation (2 Sam 19:27; 1 Kgs 13:13, 23) such as putting on a headgear or a bridle (*meteg*, Prov 26:3).

Donkeys as Booty and Tribute
The Egyptians, who kept detailed records of booty and tribute, seem to have completely disregarded seizing the donkey in spite of their awareness of its usefulness. Only once, on the Barkal Stele from his forty-seventh year, Thutmoses III reports that in his campaign against the princes of Retenu (Canaan) and upon their surrender, he let them return home to their cities "on donkey(back), so that I might take their horses . . . " (Pritchard 1969a:238).

Figure 3.6. *White donkey with blanket saddle.* (Photograph by the author.)

Obviously, Thutmoses was more interested in riding horses than in donkeys. Amen-em-heb, who fought under Thutmoses III, recalls in his biography on the walls of his tomb in Thebes that, at the battle of the Ridge of Wan, west of Aleppo, he took as booty "70 live asses" (Pritchard 1969a:241). For the Egyptians, who went on countless military campaigns in Asia and brought home animate and inanimate treasures, the avoidance of the donkey as tribute or booty is unexplainable, especially since the camel was introduced into Egypt only at the beginning of the Persian period.[23]

The Assyrians had a completely different attitude regarding donkeys. It is indisputable that they appreciated the usefulness of the donkey and, together with other animals, brought it back from their military campaigns as tribute and booty. Returning from his campaign in Judah, Sennacherib (704–681 B.C.E.) took home both male and female prisoners and different kinds of animals, including donkeys (Pritchard 1969a:288). Esarhadon (680–669 B.C.E.) describes on Prism A his campaign in Syria-Palestine and reports that from Sidon "I drove to Assyria his teeming people which could not be counted, (also) large and small cattle and donkeys" (Pritchard 1969a:290). On Prism B he describes the same campaign saying "From Sidon I led to Assyria . . . donkeys in great quantities" (Pritchard 1969a:291). When Ashurbanipal (668–633 B.C.E.) describes his campaign against Arabia he says, "I caught them all myself in their hiding-places; countless people—male and female—donkeys, camels, large and small cattle, I led as booty to Assyria" (Pritchard 1969a:299). From his description, it is apparent that the donkeys (and other animals) were raised in Arabia because he refers to "the camel foals, the donkey foals, calves or lambs [which] were suckling many times on the mother animals, [and] they could not fill their stomachs with milk" (Pritchard 1969a:300). It is interesting to note that the Judean inventory of pack and draft animals as depicted on Sennacherib's reliefs does not include any donkeys.

Horses
The horse (*sûs*, Gen 49:17; 1 Kgs 10:29) appears in the Bible as a full-fledged domesticated animal used for riding and pulling chariots (Fig. 3.7). But what do we know about the history of the horse (*Equus caballus*) and how it reached this position? As can be seen from cave paintings in France and Spain, the horse has been part of human culture since the Magdalenian phase of the Upper Paleolithic (35,000–18,000 B.C.E.). Since then, for about 15,000 years,

Figure 3.7. *Horse.* (From Bodenheimer 1935:pl. XIII.)

wild equids were hunted to provide human populations with meat and hides (Simpson 1951:24).

Throughout human history, horses were used for meat, milk, hide, draft, and riding. According to Clutton-Brock, horse and donkey domestication took place about 6,000 years ago (1992:38, 53).[24] Horse remains, some of which meet the dimensions of the present-day Przewalski's horse, appear throughout Europe at sites from the Neolithic period (Clutton-Brock 1992:54-5). Horse bones from the Chalcolithic Negev sites of Shiqmim and Gerar suggest that the horse was already domesticated by that time (Clutton-Brock 1992:55–6).[25] Recently, upon further consideration of the evidence, Grigson has suggested that the presence in the Negev of fourth- and third-millennia horse remains "raises the possibility that vehicles with spoked wheels were already being utilized there" (Grigson 1993:653). Davis, who identified horse bones in the Early Bronze Age (ca. 2280–2080 B.C.E.) strata at Arad in the northeastern Negev, cannot state with any degree of certainty that these were domestic horses. However, since wild horse remains have not been found in Israel in Mesolithic or Neolithic sites, he considers it likely that the Arad horse was already domestic (Davis 1987:164).

The spread of the horse in the ancient Near East is amply documented. Middle Bronze Age (2000–1800 B.C.E.) remains from Tal-e Malyan show the expansion of the domestic horse into southern Iran (Davis 1987:56). As an animal of speed, the value of harnessing the horse as an engine of war was probably clear. The appearance of the chariot in the Near East in the mid-second millennium B.C.E. is associated with newcomers such as Kassites, Hittites, and Hurrians, peoples who dominated Western Asia (Davis 1987:164). The chariot, which during the New Kingdom became a well-known symbol of pharaonic might, was probably introduced into Egypt during the Hyksos period (ca. 1650–1550 B.C.E.).

Harnessing the horse in the service of human culture was not without certain difficulties, however. Horses are very delicate animals that need tender, loving care. Present-day practices can serve as an example of how much care a horse requires to be in top shape (see Rose and Hodgson 1993). That horses in antiquity were highly pampered is exemplified by the grooming and the decorations with which they were lavished, as depicted in artistic representations and in archaeological finds (Pritchard 1969b:fig. 27; Elat 1977:pl. 7; Anthony 1984:7, 20–1, 23).

While it would have been obvious to ancient people's that horse's hooves (*ʿiqqbēy sûs*, Judg 5:22; *parsôt ʾabbîrāyw*, Jer 47:3) needed protection, it is not clear when horseshoes originated. The Roman hippo-sandal, which was made of metal and was attached to the bottom of the hoof with leather straps, could have been the earliest horseshoe; but it is possible that even earlier, during Iron Age II, similar attachments to the hooves of threshing animals (Borowski 1987:65) were developed for horses.

Chariotry and Horseback Riding

Simpson suggests that early domestic horses were small in stature and were more commonly used for pulling chariots than for riding (Simpson 1951:34). This does not seem reasonable, however, since pulling a cart or a chariot requires more strength, which is in direct relationship to body mass. Furthermore, Simpson claims that "horses resembling Arabians were probably among the first to be produced by domestication" (1951:36);[26] and although these are not big animals, they are well-known as riding horses. According to Anthony, horseback riding started in the Ukrainian steppes ca. 3800–3500 B.C.E., well before the introduction of carts at about 3500 B.C.E. (Anthony 1991:265–7). During that time, the horse was utilized by herding societies for moving and transporting materials from one locale to another.

When and why horseback riding arrived in the Near East has been a subject of considerable study. Horses were not widely adopted in the Near East before ca. 2300 B.C.E. (Anthony 1991:273). The "Song of the Sea" in Exodus 15:1, which is considered to be an early composition (Cross 1973:112–44), describes the Israelites' victory over the Egyptians at the Sea of Reeds (*Yam Sûp*), saying "horse and rider (*sûs wĕrokbô*) he has hurled into the sea" (see also Exod 15:21). This reference suggests that horseback riding was known in the Near East at least as early as the thirteenth century B.C.E. Schulman argues that riding horses was already practiced in Egyptian military circles for scouting purposes during the New Kingdom (Schulman 1957). He also points out that Egyptian art from the time of Seti I and Rameses II depicts mounted Asiatics, Syrians, and Hittites, who were probably also serving as mounted scouts (Schulman 1957:267–8). Several examples show the horses with reins and harness and the riders on a cloth-saddle, a few on a side-saddle. Most of these are in a military context, strongly suggesting that they are the earliest known examples of cavalry (Schulman 1957:271).

The employment of cavalry and chariotry became prevalent in Canaan even before 1000 B.C.E.[27] That the Egyptians used both is well known and the biblical record alludes to it (Josh 24:6). The Bible suggests that the Canaanites used chariotry to their disadvantage against the Israelites (Judg 4; 5:28). Also the Philistines are said to have used chariots in large numbers (1 Sam 13:5; 2 Sam 1:6). Even before its formation, the Israelite monarchy was expected to make use of these military forces (1 Sam 8:11).[28] According to the biblical account, King David did not know what to do with the chariot horses he captured from Hadadezer the Rehobite, king of Zobah, and he took most of them out of commission by hamstringing them (2 Sam 8:4).[29] But Solomon seems to have used horses for chariots and cavalry (*pārāšîm*, 1 Kgs 5:6; 9:19). He built chariot cities (1 Kgs 10:26), and was involved in horse and chariot trading serving as a middleman between Egypt, Que, the Hittites, and the Arameans (1 Kgs 10:28–29; 2 Chr 1:17). As Watkins observes, the use of chariotry "became a means of displaying the highest level of military investment" because it required large teams of trained soldiers, auxiliary forces engaged in the building and maintenance of the chariotry, weaponry, and horses (Watkins 1989:28). Solomon was no exception, because in addition to other functions, his regional governors "provided also barley and straw, each according to his duty, for the horses and chariot-horses where it was required" (1 Kgs 4:28).

Using horses for chariotry (*rekeš*, 1 Kgs 5:8; Micah 1:13) and cavalry (*rekeb*, 2 Kgs 7:14)[30] in the military became very popular between the tenth and seventh centuries B.C.E.[31] The biblical record devotes much space to descriptions of such activities. The Arameans were well known for their mounted forces. David destroyed a large number of these forces (2 Sam 10:18); and, during their conflict with David, the Ammonites hired Aramean mounted forces (1 Chr 19:6–7, 18).[32] Shalmaneser III's (858–824 B.C.E.) annalistic reports testify to the power of the Aramean and Israelite chariotry and cavalry. One such report states in part, "1,200 chariots, 1,200 cavalrymen, 20,000 foot soldiers of Adad-'idri (i.e. Hadadezer) of Damascus, . . . 700 chariots, 700 cavalrymen, 10,000 foot soldiers of Irhuleni from Hamath, 2,000 chariots, 10,000 foot soldiers of Ahab, the Israelite . . . " (Pritchard 1969a:278–9).[33] The biblical records also show that the kings of Judah and Israel employed cavalry and chariotry in many of their wars (2 Chr 21:9; 1 Kgs 16:9). The prophets describe the mounted forces of the Judahite kings (Jer 17:25;

22:4), as well as those of the enemies (Isa 22:6–7; 31:1; Jer 46:9; Ezek 26:7). That the kings of Judah possessed chariots is also illustrated on Sennacherib's palace reliefs (Fig. 3.8), where a chariot is seen taken away from Lachish by the Assyrians as booty (Ussishkin 1982:84–5).[34]

While the Egyptians used chariotry on the battlefield to their advantage, as recorded by Thutmoses III after the Battle of Megiddo (1485 B.C.E.) and by Rameses II after the Battle of Qadesh (ca. 1280 B.C.E.), the Assyrians developed the use of cavalry[35] as well as non-military mounted messengers. The Assyrians incorporated prisoners of war in their army, who were skilled in the arts of chariotry and cavalry (Elat 1977:82; Wiseman 1989:41). Thus, after con-

Figure 3.8. *Sennacherib's reliefs illustrating the fall of Lachish depicting some of the Israelite animal and vehicle inventories.* (Ussishkin 1982:83.)

quering Samaria in 722 B.C.E., Sargon II incorporated from the prisoners a contingent of fifty chariots in his royal corps (Pritchard 1969a:284–5).

The extensive use of horses for chariotry and cavalry demanded large numbers of animals and the Assyrians procured them through breeding, booty, tribute, and import.[36] The reliefs on the gates of Balawat show herds of horses brought as tribute (Elat 1977:78 and pl. 5:1–2). From official records, it appears that a certain type listed as Kusean horses (possibly Nubian, imported through Eretz Yisrael) were kept mostly for breeding (Wiseman 1989:44).[37] If the source of this type horse is correctly identified, it could also have been a source for horses employed in the Israelite and Judahite military. Solomon, who had a large force of cavalry and chariotry, acquired horses by trade with Egypt (1 Kgs 10:28–29; 2 Chr 1:17); Isaiah was also aware of the fact that Egypt was a source of horses (Isa 31:1, 3).

In the seventh century B.C.E., the Assyrians mastered the art of shooting arrows from a galloping horse.[39] Horses were ridden bareback without stirrups, which were a later development. Assyrian bas-reliefs dated from the seventh century B.C.E. show that by then horses were ridden by the nobility for hunting and by ordinary soldiers in battle.[39] Donkeys or mules are not depicted in Assyrian reliefs probably because they were not used as riding animals in war or hunting expeditions.

At first, riders used only bridle and bit, without any saddle or stirrups.[40] Many early representations show that the saddle was preceded by a cloth placed on the back of the horse. A skeleton of a mare buried in Thebes in its own coffin with a saddle-cloth on its back, dated to ca. 1430–1400 B.C.E., is a good illustration of this custom. The earliest use of a bit is illustrated by the remains of a horse found at the Egyptian fortress in Buhen, dated to as early as 1675 B.C.E., some of whose teeth show wear that could only have been caused by a bit.[41] The Hyksos (ca. 1650–1550 B.C.E.) are credited with the introduction into Egypt of horses and chariots for war and hunting (Clutton-Brock 1992:80–84). This is confirmed by zooarchaeological finds and by the earliest appearance of horses and chariots in Egyptian written records dated to this period (Pritchard 1969a:233–4).

Stabling Horses

How horses were housed raises questions for any ancient period or location, and this is also true during the Iron Age in Eretz Yisrael. Biblical and extra-biblical sources mention horses used for military purposes in Canaan, which later became Eretz Yisrael. However, with the exception of the Kusean horses mentioned earlier, neither the Bible nor any other source gives a detailed description of the horses prevalent at that time.[42] This is especially important when trying to define the function of the structures labeled "The Megiddo Stables"[43] and the other tripartite buildings.[44] One of the main arguments used by those opposing the identification of these structures as stables is that the side rooms, where the horses were supposed to have been stabled, are too narrow and could not house horses of the size with which we are familiar.[45] This problem might be resolved if the assumption concerning the size of the horses could be proven wrong, namely that small horses were stabled in these buildings.

One such possibility is Przewalski's horse (*Equus ferus przewalskii* or *Equus przewalskii* Poliakoff 1881) which was first described in 1876 and named in 1881. Although at first considered a feral, or even an hybrid pony, it was soon realized that it was a wild Mongolian equid (Clutton-Brock 1992:30–3).[46] A study by Mohr strongly suggests that Przewalski's horse is definitely an ancient breed (1971) and was probably one of the ancestors of the domesticated horse (Groves 1986:18–20).

Przewalski's horse is a small, heavy-set animal with an upright crest stretching from the top of its head to just before the shoulders. Its color is reddish-brown, somewhat lighter in the summer, with a dark median back line and shoulder line, and a dark crest, tail, and lower limbs. In the winter, its fur gets longer and lighter, and the back and shoulder stripes are absent. Once it was common all over the plains of Eurasia, but today it is limited to an area around the Altai Mountains in Siberia and to western Mongolia (Sanderson 1955:222, pl 136). An early depiction of this horse appears on a shell plaque from Susa, dated to about 2500 B.C.E., showing a small, stocky animal with a long tail (Aynard 1972:54).

Another small horse discovered recently (1964–65) in western Iran, known as the Caspian Miniature Horse (Firouz 1972), adds support to the possibility that horses of stature smaller than the normal size of present-day horses were used in antiquity for a variety of functions, including riding and drafting chariots. Illustrations of small horses used for such functions appear in the reliefs at Persepolis (5th century B.C.E.) and on a royal seal of Darius (Firouz 1972:pls. 8, 10; Clutton-Brock 1981:fig. 8.6).[47]

With the discovery of these small horses, certain questions come to mind. Is it possible that horses used as draft animals for carts, wagons, and war chariots were of one breed while another breed was used for riding? Could it be that the taller Arabian horse was used differently than was the smaller Przewalski's horse-type?[48]

In summarizing the biblical treatment of the horse, it can safely be said that the animal functioned only as a military machine and not as a power source for agricultural or other mundane tasks. Although the date of the Book of Job is later than the period under discussion here, the description of the horse in Job 39:19–25 is very characteristic of its role in daily life, namely as a military instrument.

Do you give the horse his strength?
Have you clothed his neck with a mane?
Do you make him quiver like a locust's wings,
when his shrill neighing strikes terror?
He shows his mettle as he paws and prances;
in his might he charges the armoured line.
He scorns alarms and knows no dismay;
he does not shy away before the sword.
The quiver rattles at his side,
the spear and sabre flash.
Trembling with eagerness, he devours the ground
and when the trumpet sounds there is no holding him;
at the trumpet-call he cries "Aha!"
and from afar he scents the battle,
the shouting of the captains, and the war cries.

In the realm of Israelite cultic practices, archaeological evidence such as clay figurines of horses with or without riders and the biblical account in 2 Kgs 23:11 suggest that the horse was somehow connected with sun worship, possibly in a symbolic or a real, physical way (Taylor 1993: 58–9, 176–8, 262–3).

Horses and Chariots in the Ancient Near East
The earliest mention of a chariot and horses used in Egypt is in a text "The Expulsion of the Hyksos," where Pharaoh is described as having a chariot

(Pritchard 1969a:233–4). The Hyksos introduced horses and chariots in Egypt, and the Egyptians adopted them.

In his reports on the Asiatic campaigns, Thutmoses III continues to describe the major role played by horses (Pritchard 1969a:235, 236). His enemies' horses and chariots were captured, and some of the booty he carried off from Megiddo includes "2,041 horses, 191 foals, 6 stallions, . . . colts" and a total of 924 chariots (Pritchard 1969a:237). The Barkal Stele summarizing Thutmoses's achievements states, "tribute: gold and silver, all their horses which were with them, their great chariots of gold and silver . . . " (Pritchard 1969a:238). At every one of his campaigns, he took booty that included horses and chariots (Pritchard 1969a:239, 241).

When Amen-hotep II (1447–1421 B.C.E.) reports on his Asiatic campaigns, he claims that near the Orontes river he captured two princes and six maryanu with "their chariots, their teams, and their weapons of warfare" and on his return to Memphis, the list of booty includes "horses: 820; chariots: 730 . . . " (Pritchard 1969a:246). More horses and chariots were taken as booty at the battles of Aphek and at Iteren and Migdol-yen (unknown sites) (Pritchard 1969a:247). On his return to Memphis the list of booty includes "chariots of silver and gold: 60; painted chariots of wood: 1,032 . . . " (Pritchard 1969a:247).

The Assyrians also replenished their chariotry and cavalry needs by military campaigns. Tiglath-Pileser I (1114–1076 B.C.E.) brought back from his campaign in the Levant, where he fought and defeated thirty kings, "horses, broken to the yoke" (Pritchard 1969a:275). Ashurnasirpal II (883–859 B.C.E.), on his expedition to Carchemish and the Lebanon "took over the chariot (-corps), the cavalry (and) the infantry of Carchemish" (Pritchard 1969a:275, 276). Shalmaneser III (858–824 B.C.E.), in his so-called Monolith Inscription states about his first year, "I took away from him many chariots (and) horses broken to the yoke" (Pritchard 1969a:277). The large numbers of mounted forces involved in these conflicts is evident from Shalmaneser III's sixth-year statement in the Monolith Inscription.

He [Irhuleni from Hamath] brought along to help him
1,200 chariots, 1,200 cavalrymen, 20,000 foot soldiers of
Adad-'idri (i.e. Hadadezer) of Damascus (Imerisu), 700
chariots, 700 cavalrymen, 10,000 foot soldiers of Irhuleni

> *of Hamath, 2,000 chariots, 10,000 foot soldiers of Ahab,*
> *the Israelite (A-ha-ab-bu* matSir-'i-la-a-a), *500 soldiers*
> *from Que, 1,000 soldiers from Irqanata, 200 soldiers of*
> *Matnu-ba'lu from Arvad, 200 soldiers from Usanata, 30*
> *chariots, 1[0?],000 soldiers of Adunu-ba'lu from Shian,*
> *1,000 camel-(rider)s of Gindibu', from Arabia, [. . .],000*
> *soldiers of Ba'sa, son of Ruhubi, from Ammon—(all*
> *together) these were twelve kings . . . . Even during the*
> *battle I took from them their chariots, their horses broken*
> *to yoke (Pritch*ard 1969a:278–79, and reports from his
> other campaigns on 279–81).

All other Assyrian kings continued the tradition of depleting their defeated enemies of their equid resources. Tiglath-Pileser III (744–727 B.C.E.), in his campaigns to Syria and Palestine took, as booty, "Horses, mules (trained for) the yoke . . . " (Pritchard 1969a:282). On another campaign to Syria and Palestine, he brought back as booty "horses, mules, large and small cattle, (male) camels, female camels with their foals" (Pritchard 1969a:283).

Sargon II (721–705 B.C.E.) used prisoners to form units of chariots (Pritchard 1969a:284–85), and he brought horses as booty and tribute from Arabia (Pritchard 1969a:285–86).

Sennacherib (704–681 B.C.E.) brought back from his campaign in Judah "horses, mules, donkeys, camels, big and small cattle beyond counting and considered them booty" (Pritchard 1969a:288), and Ashurbanipal (668–633 B.C.E.), during his second campaign against Egypt and Nubia, carried away booty from Thebes including "fine horses" (Pritchard 1969a:295, 297).

Mules

All equids can interbreed and produce offspring, however they are almost always sterile.[49] These hybrids are larger, have greater endurance, and can survive better on poor food (Clutton-Brock 1992:42). The descendent of a male donkey (jack or jackass) and a female horse is a mule, known by its scientific name *Equus asinus* x *Equus caballus*. A hinny or jennet is the progeny of a male horse and a female donkey (jenny); it is known by its scientific name *Equus caballus* x *Equus asinus*. Mules and hinnies (biblical *pered*, Zech 14:15; Ps 32:9; and *pirdâh*, 1 Kgs 1:33) usually have consistent characteristics that combine the looks of their parents. The mule is donkey-like in appearance, with a more horse-like body and relatively long ears; its tail is ass-like, and its legs are fine-boned with small hooves. The more horse-like hinny is small with the body of a donkey; its head is lighter, the ears shorter, and the tail fuller. This follows the general rule that the head and tail of the hybrid inherit the characteristics of the sire. One exception to the donkey-like character of the mule is that it lacks the white belly of its male parent (Lydekker 1912:235; Clutton-Brock 1992:44–45). The mule is generally stronger and more robust than a hinny, and at least since the Roman period, this hybrid has been bred more frequently than the hinny. Clutton-Brock suggests that, during Sumerian times, more hinnies were bred because a rare imported stallion could produce a whole herd of hybrids in the time it would take one mare to produce a mule (Clutton-Brock 1992:45).

The Sumerians may have been the first to breed hybrids between the domestic donkey (descended from the African wild ass, *Equus africanus*) and the wild Asiatic ass, *Equus hemionus.* The most important evidence for breeding domestic donkeys with wild onagers is found in numerous cuneiform clay tablets coming from Sumerian sites in Mesopotamia. Different cuneiform signs have been transcribed as meaning wild, domestic, and hybrid equids. The Sumerian texts reveal that donkey-onager crossbreeding was done in the earlier period, around 2800 B.C.E., but that later in the third millennium B.C.E., using the onager was discontinued and mule-breeding between donkeys and horses became the common practice (Clutton-Brock 1992:43–4). Through experimentation, it was found that even stronger hybrids could be obtained by crossing a male donkey with a female horse, and this became the accepted method of producing the most powerful and resilient pack and draft animal for peace and war purposes (Clutton-Brock 1992:14).[50] The process that produced true mules could only have been achieved after the domestication of the horse and the wide availability of mares.[51] Once it was developed, since it could not reproduce itself, the mule became an expensive item. Hittite records show that while the price of an ox was equal to ten sheep and that of a horse to twenty sheep, the price of a mule equaled sixty sheep (one sheep = one shekel) (Dent 1972:62).

In the third millenium B.C.E., lack of reproductive ability priced mules higher than other equids. While naturally reproductive equids were priced from four to seven silver shekels, the price of a mule ranged between twenty to thirty shekels (Zarins 1986:185). In Ebla, the highest price of a mule was five mina or 300 shekels (Zarins 1986:187). Remains of *E. caballus* x *E. asinus* in South Turkey and North Syria indicate "that in the Bronze Age in North Syria mules and/or hinnies were deliberately bred on a large scale to meet the needs of developing trade" (Buitenhuis 1991).

Artistic representations from Egypt dated to the 18th Dynasty show the use of mules (hinnies) as draft animals for two-wheeled chariots (Clutton-Brock 1981:fig. 9.9; Clutton-Brock 1992:86). Dent suggests that these animals might be either onagers or mules by onager stallions (Dent 1972:36). Since onagers were not domesticated, the latter suggestion seems to be the more reasonable. While domestic donkeys are rarely depicted in Assyrian reliefs, some show mules as pack animals, a relief from Ashurbanipal's palace in Nineveh dated to ca. 645 B.C.E., for example, shows a mule carrying a woman and a child (Clutton-Brock 1992:92, fig. 6.14). Another Assyrian relief from the same palace depicts a male mule carrying hunting gear.

Mules in Ancient Israel

Biblical references to the mule are very few. It appears at least three times as a working animal in the general animal inventory (1 Kgs 18:5; Zech. 14:15; Ps 32:9). Moreover, in biblical circles the mule was considered an animal of high status that was used by people in high standing and authority. The king's sons rode mules (2 Sam 13:29); during his rebellion against David, Absalom rode a mule when he fled away; and during his coronation, Solomon rode the king's hinny (*pirdat hammelek*, 1 Kgs 1:33, 38, 44).[52] The documented use of the mule by Israelite royalty and nobility raises some questions concerning the prohibition against crossbreeding (Lev 19:19) in Israelite society. What was the source of these hybrid animals? Is it possible that the Israelites did not adhere to this prohibition, or was this prohibition introduced some time later? Was it possible that the prohibition was observed but that hybrid animals were allowed to be used if they were obtained from outside sources?[53] Indeed, it is stated in 1 Kgs 10:25 and 2 Chronicles 9:24 that mules were brought to Solomon as tribute and in return for his words of wisdom.

Being a strong animal, the mule was used not only for riding but also for physical work—hauling soil (2 Kgs 5:17), carrying foodstuffs (1 Chr 12:40[41]), and as draft animal harnessed to a cart or wagon (Isa 66:20).[54]

Mules continued serving the Judahites as pack and draft animals upon their return from exile in Mesopotamia.

The whole assembled people numbered forty-two thousand three hundred and sixty, apart from their slaves, male and female, of whom there were seven thousand three hundred and seventy-seven; and they had two hundred male and female singers. Their horses numbered seven hundred and thirty-six, their mules two hundred and forty-five, their camels four hundred and forty-five, and their donkeys six thousand seven hundred and twenty. (Ezra 2:64–67)

Mules in the Ancient Near East

Mules began to appear in booty and tribute lists after Tiglath-Pileser III's (744–727 B.C.E.) campaign in Syria and Palestine. Mules were included in his booty list after both this and a later campaign in that region. Sennacherib, after his campaign in Judah, also returned with mules and other animals, which he considered booty. While Assyrian military prowess replenished and enriched their own animal inventories, it seems on the surface that this would have depleted their enemies' animal resources. However, the fact that they could return periodically and confiscate additional animals strongly suggests that the Assyrians had a policy through which they enabled local populations to breeding animals for future appropriation.

~

While the horse served a very narrow role in the cultures of the Ancient Near East as a prestigious draft and riding animal, the donkey and the mule performed many more functions as working and riding animals that were used by common people as well as the nobility. All of these animals were valued for their abilities and this was reflected in the fact that booty and tribute frequently included large numbers of them. Nevertheless, the horse must have had a particularly respected position in Israel judging from the number of Iron Age clay figurines depicting horses found at many archaeological sites (Holland 1977). The context of many of these figurines suggests that the horse had a central place in the cult.

OTHER DRAFT AND PACK ANIMALS

Camels

The camel (*gāmāl*, Gen 24:64) belongs to the Cameloid family (*Camelus dromedarius*, one-humped camel; *Camelus bactrianus*, two-humped camel). Originating in South America, the family also includes the llama (*Lama glama*), alpaca (*L. pacos*), guanaco (*L. guanicoe*), and vicuña (*Vicugna vicugna*) (Clutton-Brock 1981:123; Yagil 1993:30). The exact time, purpose, and place of the domestication of the camel are still being debated. According to Davis, the camel was domesticated some time during the third millennium B.C.E. (1987:127, 166), but it did not reach northwest Arabia and Syria-Palestine before the end of the second millennium (Bulliet 1975:36, 64). Bulliet claims that the camel was originally domesticated in southern Arabia for the use of its milk, and from there, around 2500–1500 B.C.E., was introduced at into Somalia and Socotra (1975:45, 56).

However, Clutton-Brock proposes that the camel was first domesticated for use as a pack animal and only later certain societies learned to use it for other by-products and services (1981:129). Furthermore, there is a suggestion that the dromedary (Fig. 3.9), which is better adapted to hot climates than the Bactrian camel, was developed in Arabia from the two-humped camel especially to withstand heat (Bulliet 1975:34–5).[55] This could have been related to the rise of the overland incense trade that was developed by Semites around 2000 B.C.E. (Bulliet 1975:58). "They [Joseph's brothers] had sat down to eat when, looking up they saw an Ishmaelite caravan (*'orḥat yišmĕ'ē'lîm*) coming from Gilead on the way down to Egypt, with camels carrying gum tragacanth and balm and myrrh" (Gen. 37:25–28). While the accuracy and date of this biblical episode are in dispute, its mention might reflect a long-standing tradition in which the Ishmaelites who appear in the Joseph story as traders in incense are the descendants of the original incense merchants. Davis, however, proposes that both types of camels must have been independently domesticated, the dromedary perhaps in southern Arabia—possibly in Hadarmaut, where it may initially have been exploited for its milk—and the Bactrian in Asia—perhaps in Persia (1987:166). Uerpmann, after reviewing the available data, concludes that the wild dromedary was at home in Arabia as late as the third millennium B.C.E. and might have reached as far west as

Figure 3.9. *Camel.* (From Bodenheimer 1935:pl. xiii.)

Arad (1987:52). This excludes the possibility of the Bactrian camel being the progenitor of the dromedary.

The question of the time and place of camel domestication has not been resolved yet, and this is due in part to the scarcity of evidence. According to Köhler, this scant evidence is the result of a concentration of archaeological work in populated areas where the camel did not live naturally. In addition, camels were rarely eaten, and their bones are thus not as readily available as those of other animals (Köhler 1984:201).

The domestication and widespread use of the camel brought further improvements in trade in the second millennium B.C.E. and were probably very important in the development of a new lifestyle of desert nomadism in the Near East (Köhler 1984; Davis 1987). This lifestyle was possible thanks to the camel's anatomy. The camel's hump, which serves for fat storage, was probably developed as a body-heater; for water storage, the animal has several sac-shaped extensions in its stomach where liquid can be retained for a long period (Sanderson 1955:248; Clutton-Brock 1981:123). That this resource can be used in cases of emergency was reported by Ashurbanipal (668–631 B.C.E.) after one of his campaigns, in which he forced his thirsty enemies to "slit open camels, their (only) means of transportation, drinking blood and filthy water against their thirst" (Pritchard 1969a:299).

Textual references to the camel dated to before the end of the second millennium B.C.E., as well as finds of skeletal remains from the period in the Near East are rare. That led Albright to conclude that the camel could not have been domesticated before the end of the twelfth century B.C.E. and that references to camels in Genesis and Exodus are anachronistic (1949:207). Albright maintained that the camel's early appearance in Bible stories in Genesis 12:14–16; 24:10–67; 31:17–35; 37:25 is out of place. Accordingly, its first appearance in the Bible at the proper historical time is during the period of the Settlement, as described in Gideon's encounter with the Midianites in Judges 6–7.

But Bulliet maintains that close examination of the domestication process and the evolution of relatively late implements indicate that the camel was actually domesticated in southern Arabia long before the year 1100 B.C.E. The use of the animals did not penetrate Arabia's northern regions until the period suggested by Albright, although awareness of it may have (Bulliet 1975:36, 38). While it is tempting to agree with Albright in dating the beginning of camel use in Syria-Palestine to the last two centuries of the second millennium, there is physical evidence that refutes his argument and cannot be avoided. Possible evidence from the Near East shows camel domestication and use, although not wide-spread, in 2500–1400 B.C.E. (Bulliet 1975:58–64). This simply means that in the nineteenth and eighteenth centuries B.C.E., when, according to Albright, the Patriarchs might have lived, camels were already known in small numbers in the northwestern corner of the Arabian desert, where the western Arabian trade route branched out to go to Egypt or further into Syria. It is quite possible that local tribes in the area owned a few of the animals, perhaps as articles of prestige, without being heavily involved in breeding them (Bulliet 1975:65). Thus, the presence of camels in the Patriarchal stories can be defended, and the story can be treated as primary evidence of camel use without disputing Albright's contention that camel-breeding nomads did not exist in Syria and northern Arabia until later (Bulliet 1975:67).[56]

Although the camel was most likely first domesticated for its milk,[57] other uses were found later, including using its hair,[58] meat, and dung, and utilizing it as a riding and pack animal (Fig. 3.10). What makes the camel a good pack animal is that it can carry twice the load of a donkey, moves faster, and needs less feeding and watering (Davis 1987:166). However, the camel's anatomy requires a special harness with straps, known as withers-straps, that go between or around the humps or are supported by a saddle over one hump (Bulliet 1975:182). Like the ox, which has a usable skeletal structure in the long vertebrae joining the neck to the body and thus requires a special harness, the camel, whose hump is without skeletal support, also needs a specially developed harness (Bulliet 1975:178, 182). Once becoming a pack animal, the camel was used in many Near Eastern communities, including carrying equipment during the *raḥil*.[59]

The reason camel breeding finally caught on in Syria-Palestine seems to lie in saddle development. When the camel's function changed from a milk producing to a pack animal, it created a reason to maintain male camels,[60]

Figure 3.10. *Camel with a riding saddle.* (Photograph by the author.)

and this was bolstered by the development of the behind-the-hump saddle (*kar haggāmāl*, Gen 31:34), which was used for riding pack animals (Bulliet 1975:68–76). This is illustrated in a ninth-century B.C.E. relief from Tell Halaf (Fig. 3.11) showing a camel ridden by a man seated on a box-like saddle and carrying a stick or sword in his hand (Pritchard 1969b:fig. 188).

As a herd animal, the camel became known in Syria-Palestine only after control of the incense trade had passed to the Semites, who then had a good use for the animal. The pattern of camel use adopted by these herders was, naturally, the new one in which its utilization as a pack and riding animal[61] played a major role, rather than the older pattern of being entirely subsistence related (2 Kgs 8:9). Its by-products continued to be used, however, especially dung and hair, which, together with black wool, was used for weaving tents and nets (Forbes 1956:277).[62]

The Camel in the Bible

Whether the camel was domesticated first for its milk or meat, there is a clear biblical prohibition against consuming its meat (Lev 11:4; Deut 14:7).[63] As a herd animal, the camel appeared earlier on the western side of the Syrian desert than on its eastern side (Bulliet 1975:77), and this might be reflected in the appearance from Gilead of the Ishmaelite camel caravan in the Joseph story (Gen 37:25). Another biblical tradition relates that, during the Settlement period, the Midianites and Amalekites inflicted heavy suffering upon the Israelites by invading their territory and camping with their animals on sown land until subdued by Gideon. As related in Judges 6:5 "they and their camels were past counting." That the Midianites and Amalekites owned camels did not strike the storyteller as being extraordinary, but the large numbers did.

The camel, as an Amalekite pack animal, is mentioned in Samuel's command to Saul: "Go now, fall upon the Amalekites, destroy them, and put their property under ban. Spare no one; put them all to death, men and women, children and babes in arms, herds and flocks, camels and donkeys" (1 Sam 15:3). Camels were used for riding (*rekeb gāmāl*, Isa 21:7) and were given as royal presents, as when Ben-Hadad gave Elisha forty loaded camels (2 Kgs 8:9). The Bible records that David had large herds of camels under the charge

Figure 3.11. *Camel ridden by a man seated on a box-like saddle. Relief from Tell Halaf.* (Pritchard 1969:fig. 188.)

of an individual from Arabia, "Obil the Ishmaelite was in charge of the camels" (1 Chr 27:30). Like other domesticated animals, the camel was susceptible to diseases, as mentioned in Zechariah 14:15, "plague will also be the fate of horse and mule, camel and donkey. . . . "

From the biblical record, it is evident that the camel was used primarily as a pack and riding animal, the same role it played during that period throughout the Near East. Bone remains from several sites in Israel support the idea that the camel appeared in this region at the end of the Settlement period and became integrated in the overland transportation framework (Hakker-Orion 1984). Surprisingly, no camel bones were identified at Iron Age I strata in Beersheba (Hellwing 1984), a site on the border of the Negev Desert. However, since donkey bones were found there, it is very possible that donkeys were still used for transport in this arid zone at that period and were not yet replaced by camels. There is no reference that can be construed that the camel played a role in performing any agricultural chores such as plowing or threshing.[64]

Camels in the Ancient Near East
Ancient Near Eastern documents, especially Assyrian annals, mention camels in different contexts. From written comments and artistic representations, it appears that the Arabs used camels in their fighting forces (Fig. 3.12). Royal reliefs from the time of Ashurbanipal (668–631 B.C.E.) depict the Arabs with their fighting camelry being chased by the victorious Assyrian cavalry (Pritchard 1969b:fig. 63; Bulliet 1975:83, fig. 34). Written references make it quite clear that the Assyrian's primary camel source was Arabia, with Syria-Palestine the secondary source. The sixth-year Monolith Inscription of Shalmaneser III (858–824 B.C.E.) mentions "1,000 camel-(rider)s of Gindibu', from Arabia" (Pritchard 1969a:279).[65] However, Bactrian camels were also welcomed from other parts of the Near East. His Black Obelisk shows two-humped camels (Bulliet 1975:fig. 70) and refers to "the tribute of the country Musri; I received from him camels whose backs were doubled" (Pritchard 1969a:281). Two-humped camels (Bulliet 1975:fig. 71) and camels with a south Arabian saddle were depicted by Shalmaneser on the gates of Balawat (Bulliet 1975:74, fig. 28).

Another Assyrian king, Tiglath-Pileser III (744–727 B.C.E.), mentions in his annals of a campaign to Syria and Palestine in an unknown year that the

Figure 3.12. *Ashurbanipal's war against Arab camelry, from Kuyunjik.*
(Frankfort 1954:102.)

booty included "(male) camels, female camels with their foals" (Pritchard 1969a:283). After another campaign in Syria, Palestine, and Arabia in an unknown year, he received from Arabia male and female camels as tribute (Pritchard 1969a:283) and he states that he killed 30,000 camels and 20,000 head of cattle that belonged to Samsi, queen of Arabia (Pritchard 1969a:284). His camel acquisition through military campaigns is depicted on a relief from his palace in Nimrud showing captive camels being led by a woman (Pritchard 1969b:fig. 187).

Sargon II (721–705 B.C.E.), in the annals of his first year states that he received "from Samsi, queen of Arabia, (and) It'amar the Sabaean, gold in dust-form, horses (and) camels" (Pritchard 1969a:285). His seventh year annals also state that he received "as their presents . . . horses (and) camels" (Pritchard 1969a:286). During his reign, camels were used in the Assyrian military as pack animals (Bulliet 1975:78).

On the Oriental Institute Prism, Sennacherib (704–681 B.C.E.) makes the following claim after his campaign in Judah: "I drove out (of them) 200,150 people, young and old, male and female, horses, mules, donkeys, camels, big and small cattle beyond counting and considered them booty" (Pritchard

1969a:288). That the Judeans had camels is illustrated on Sennacherib's relief (Fig. 3.8) which shows Judean refugees from Lachish leaving the besieged city with a loaded camel (Bulliet 1975:80, fig. 32; Ussishkin 1982:84–85). Camel bones were found in an Iron Age stratum at Lachish (Bates 1953), indeed verifying the presence of this animal at the site during Sennacherib's time.

His successor, Esarhadon (680–669 B.C.E.), in the description of his campaign against the Arabs and Egypt on Prism B, boasts that

> *As an additional tribute, I imposed upon him [Hazail] (the payment of) 65 camels (and) 10 foals (more than) before. When fate carried Hazail away, I set Iata', his son, upon his throne and assessed upon him an additional tribute of 10 minas of gold (and) 1,000 birûti-stones, 50 camels, 100 kunzu-bags [some kind of leather bag] with aromatic matter (more than) his father (paid). (*Pritchard 1969a:292)[66]

Following his campaign against Arabia, Asshurbanipal (668–631 B.C.E.) claimed that he "took as booty from them countless prisoners, donkeys, camels, and small cattle" (Pritchard 1969a:299). His detailed description of the campaign assigns the camel an important role in desert crossing and as booty.

> *I ordered soldiers to stand on guard . . . where there were cisterns or water in the springs, thus refusing them (the access to the) water (supply) which (alone) could keep them alive. I (thus) made water to be very rare for their lips, (and many) perished of parching thirst. The others slit open camels, their (only) means of transportation, drinking blood and filthy water against their thirst . . . . I caught them all myself in their hiding-places; countless people— male and female—donkeys, camels, large and small cattle, I led as booty to Assyria. . . . I formed flocks and distributed camels as if they be sheep, dividing (them) up to all inhabitants of Assyria. Camels were bought within my country for less than one shekel of silver in the market place.* (Pritchard 1969a:299)[67]

Echoes of these campaigns may be found in Assyrian palaces. A relief from Kuyunjik attributed to Ashurbanipal (668–633 B.C.E.) depicts two very thin dromedaries with their ribs showing and humps almost completely gone. Since they are not bactrian camels, the relief may illustrate one of his Arabian campaigns rather than a campaign in Elam. The other animals in the relief are fat and healthy, suggesting that the camels belonged to the besieged people (Pritchard 1969b:fig. 170).

The Assyrian and later military campaigns against Egypt might be traced in the zooarchaeological record. About 500 dromedary bones were found at Tell Jemmeh, the majority of which belong to the Assyrian (675–600 B.C.E.) and Neo-Babylonian/Persian (ca. 600–332 B.C.E.) periods.[68] They represent only a very small proportion of the faunal remains at the site, which include cattle, equids, gazelles, pigs and birds. Camel bones represent 25% of the Assyrian faunal sample, 35–47% of the Neo-Babylonian/Persian sample, and 14% of the Hellenistic sample (Wapnish 1984:174). According to Wapnish, the camels were assembled at the site as part of the Assyrian military policies against Egypt (1996:291).[69] Camels were used for travel in the northern Sinai from Tell Jemmeh to Egypt. One such example is recorded by Esarhadon (680–669 B.C.E.). In his tenth campaign, when crossing the Sinai from Raphiah he "put [water bottles] . . . upon the camels which all the kings of Arabia had brought" (Pritchard 1969a:292). Furthermore, on the basis of bone remains, Wapnish proposes that the large numbers of animals claimed by the Assyrian kings as booty and tribute might not be "literally accurate, but they reflect the ancient economic realities well enough to deflate the charge of mere boast" (1996:292), hence the numbers of animals reported by the Assyrian kings taken as tribute and booty are not too far from the truth.

Oxen and Working Cows

While the earliest traction animals were bovines (*Bos taurus*), this use did not evolve until man's perception of the already-domesticated animals changed from a resource for meat and other by-products to a source of muscular power that could be harnessed to carry burdens across long distances and to pull plows, sleds, wheeled carts, and even war chariots (Davis 1987:162) (Fig. 3.13). Some of the earliest examples for the use of cattle power can be seen in Sumerian pictograms and cylinder seals from Uruk in southern Mesopotamia, which date to the end of the fourth millennium B.C.E. (Davis 1987:164). These depictions continued in later periods. A seal impression from Nippur (1320-1295 B.C.E.) depicts a scene in which two humped, long-horned cattle draw a plow with a seed drill (Pritchard 1969b:fig. 86). In the Royal Tombs of Ur (early third millennium B.C.E.), the ox was harnessed to the king's chariot, and to the chariots taken as booty and tribute by the Assyrian kings in the first millennium B.C.E. (Aynard 1972:52). This is well-illustrated in a bas-relief at the palace of Ashurbanipal (668–633 B.C.E.) in Nineveh that depicts a team of short-horned oxen drawing a spoked, two-wheel cart (Pritchard 1969b:fig. 167).

The ox continued to be valued for its meat and hide once it became a beast of burden, but these were utilized most efficiently only after its physical power had been exploited.

With the invention of the wheel and the introduction of the wagon, a means for harnessing the animal was needed and the yoke ('ol, Num 19:2) was designed. With the yoke harness, most likely the earliest of the harnessing devices, the point of traction is designed to be above the animal. This is very effective for bovine animals, where the yoke can pull against the arch of

Figure 3.13. *Model of oxen pulling a cart, Anatolia.*
(The Bible Lands Museum, Jerusalem.)

Figure 3.14. *Egyptian carts pulled by oxen.* (Hawks 1850:157.)

vertebrae (seventh cervical, first to fifth thoracic) joining the neck to the body or, more rarely, be tied to the horns (Bulliet 1975:177–8). A quite life-like three-dimensional wooden model of a team of oxen yoked by the horns to a plow, dated to Sixth-Eleventh Dynasty (2350–2000 B.C.E.), now at the British Museum, exemplifies this method of harnessing (Pritchard 1969b:fig. 84).

While the earliest evidence for the use of wheeled carts comes from Sumerian pictograms, it had been suggested that both the plow and the cart were developed in northern Mesopotamia ca. 3200 B.C.E. and within half a millennium spread as far as northwestern Europe. Most of the evidence for harnessing of animal power comes from material culture finds such as plow marks, depictions on bas-reliefs (Fig. 3.14), ceramic models, accessories such as bits and, in the second and first millennia B.C.E., even texts (Davis 1987:164).

Oxen and Working Cows in the Bible and the Ancient Near East
In Israelite society, the leading animal of burden was the ox (*šôr,* Deut 15:19; 22:10),[70] which is depicted in the Bible as the main pack and traction animal (Deut 22:4, 10; 25:4). Still, the collective term for cattle *bāqār* is sometimes employed to denote a team of oxen[71] used for work, as in the case of Saul plowing his field (1 Sam 11:5, 7; see also 2 Sam 24:22; 1 Kgs 19:20 and 1 Chr 13:9; 21:23).[72] Sennacherib's reliefs illustrating his capture of Lachish (701 B.C.E.) show two examples of Judahite use of oxen as draft animals. Both instances depict a team of two long-horned oxen—their masculine gender is

very clear—pulling a two-wheel cart or wagon, one of which has six and the other eight spokes in each wheel. The carts are piled up with bags, household utensils, and people (children, women), and each is led by a man holding a stick (*malmad bāqār*, Judg 3:31). The ribs of the oxen are showing (Ussishkin 1982:84–7), possibly reminding the viewer of the results of the long siege (Fig. 3.8).

Cows were also used for work, as is implied in the instructions concerning the "Red Heifer ritual" (Num 19),[73] especially where the cow is described as "one which has never borne a yoke (*'ol*)" (19:2). There are other instances where working cows are described. Cows used as draft animals are mentioned when the Ark of the Covenant was sent back to Judah from Philistia after bringing certain diseases upon the Philistines (1 Sam 6). A heifer (*'eglâh*) performing agricultural tasks is used in several metaphors, such as the one employed by Hosea saying that "Ephraim is like a heifer broken in, which loves to thresh grain; across that fair neck of hers I have laid a yoke, I have harnessed Ephraim to the pole to plough, that Jacob may harrow the land" (10:11). Another famous case of alluding to the heifer as a working animal is that of Samson's riddle. When his wife's relatives could not solve the riddle and asked her to intervene, upon presentation of the answer Samson quipped: "If you had not ploughed with my heifer, you would not have solved my riddle" (Judg 14:18).[74]

The use of cattle in performing agricultural chores is well documented in the Near East.[75] Long-horned cattle are depicted threshing and plowing in a tomb painting from the Twelfth Dynasty at Beni Hasan (Pritchard 1969b:122). Another plowing scene with long-horned cattle yoked by the horns is seen in a tomb painting from Thebes dated to the last quarter of the fifteenth century B.C.E. (Pritchard 1969b:fig. 91). Other scenes of agricultural chores from Mesopotamia are described above.

The ox was considered very valuable; its welfare was protected (Exod 23:12; Deut 5:14) and legal provisions were made to safeguard its great value (Exod 21:37; 22:9; 23:4). However, its aggressive nature was also recognized, and its control was the responsibility of the owner (Exod 21:28–29, 32, 35–36). Unlike the donkey, a first-born ox could not be redeemed (Exod 34:19; Num 18:17). This might be significant, since it is further emphasized by the fact that the ox is almost always mentioned first in the list of sacrifices, followed by the ram, sheep, and goat (Lev 4:10; 7:23; 9:4). Oxen were fattened for consumption (Prov 15:17), and fattened oxen or bulls (*merî*, Isa 11:6) were among the choice animals for sacrifice (2 Sam 6:13; 1 Kgs 1:9, 19, 25; Isa 1:11).

∽

Pack and draft animals played an important role in the Near Eastern domestic faunal inventory in general, and for the Israelites in particular. Although most of these animals were domesticated at different times for different reasons in different regions, each of them occupied a specialized niche in Ancient Israel during the Iron Age. Most common was the donkey, which was used mainly as a pack and riding animal traversing mostly the hilly and arid regions. Furthermore, to a limited degree it continued to be used in agriculture for drawing plows and threshing. The ox became the most common draft animal, pulling wagons, carts, and plows. This function was not reserved for the male of the species, and there is ample evidence that cows were also used in the same manner. Joining these draft animals were the mule and hinny, which were used for both drafting and riding. Riding hinnys, however, was mostly the purview of the nobility. The horse was mostly used in military contexts for riding and for drafting war chariots. In the latter capacity, the horse was also utilized for hunting wild animals. The camel was the most selectively used animal, and it never reached the popularity gained by the others. Its introduction in the western part of the Near East can be dated to the transition period of Late Bronze/Iron Age; it was used mostly as a riding and pack animal in the most arid regions.

While an Israelite engaged in agricultural activities would have owned a donkey or two, or an ox, a pair of oxen, or cows, an individual would not have owned a horse, and only those engaged in long-distance trade would have owned a camel. The ownership of mules and hinnies was much more limited because of their value. An individual who owned other pack and draft animals would not have invested in any of these hybrid animals.

Notes

1. The mule, a member of this group, is such a hybrid animal. For more information on the *Equidae* family see Groves 1986.

2. An attempt to domesticate zebras was successful in South Africa at the end of the nineteenth century, but was discontinued at the outbreak of the Boer War (Clutton-Brock 1992:24). It has been suggested by Zeuner (Clutton-Brock 1992:87) that the onager was domesticated by the Sumerians, but this has been disproven by several zooarchaeologists.

3. For a discussion of the early domestic equids, especially donkeys, in the Near East see Burleigh, 1986 and Burleigh, et al. 1991. Uerpmann discusses in detail the distribution of the *Equidae* family in the Ancient Near East (1987:13–40).

4. For a summary of the different theories concerning the domestication of the horse, see Groves 1986:16–20.

5. For more information on early equids in the Near East, see Meadow and Uerpmann 1986 and 1991.

6. Other biblical terms for cart or wagon are *ṣab* and *kirkārâh* (Isa 66:20).

7. For more information on the wagons, their wheels, and harnessing, see Clutton-Brock 1992:70-73 and Anthony 1984:12–13.

8. See also Num 19:2; Deut 21:3. The term *'ol* is used metaphorically to denote servitude to foreign powers (Jer 27:8, 11–2; and more) or higher authority (1 Kgs 12:10–1, 14).

9. See the depiction on the Ur Standard and discussion in Anthony 1984:20.

10. For an example of an early bit from around the time of Rameses II see Bietak 1990:101, pl. 11.

11. For a detailed interpretation of these verses in light of a discussion of the species, see Groves 1986:42.

12. For present day varieties see Dent 1972:19–24.

13. In the mastaba of Ti (Dyn V) at Saqqara, a group of eleven donkeys is seen at the trough (Michalowski 1968:fig. 246). The front donkey, on the right, has a shoulder stripe, the mark of the *E. africanus*, the wild ass that was the progenitor of the domestic donkey. On wild donkeys and their domestication see Grove 1986:24–46.

14. Although composed in a later period, in Job 39:5, the *pere'* appears in parallelism to *'ārôd,* suggesting that both terms refer to the same or a very similar animal. A good depiction of an onager is on a bronze rein ring from the tomb of Queen Pu-Abi in Ur, dated to the first half of third millennium (Aynard 1972:pl. 3). For a description of the onager and other wild asses, see Clutton-Brock 1981:92-4.

15. A second Asiatic wild ass is known as the *kiang* (Lydekker 1912:226-7) lives in the Tibetan uplands. The kiang is much larger than the onager, has larger hoofs, and is red and white in the summer but dull brown in the winter when it is covered with shaggy hair (Sanderson 1955:222). For a study of the *E. hemionus* see (Compagnoni 1978).

16. It appears from skeletal remains that, by the Early Bronze Age, the donkey was already domesticated in Palestine (Tell ed-Duweir) (Clutton-Brock, 1986:212; Groves, 1986:41).

17. Remains of *Equus africanus* have been identified in Arabia in regions similar to those of North Africa (Uerpmann 1991).

18. One of three donkey skeletons found in Egypt by Sir Flinders Petrie was recently carbon dated to 4390 ± 130 BP (Clutton-Brock 1992:65).

19. Dent suggests that the animals are onagers, but he cannot explain why the process of their domestication was interrupted or abandoned (Dent 1972:34).

20. In the publication about the figurine, the animal is referred to as mule, but its shape and size relative to the rider indicate that it is a donkey.

21. How old an *'ayir* might be is hard to determine since, at a certain age, they were pressed into service still under this term (Judg 10:4; Isa 30:6, 24).

22. In 2 Sam 19:27, the masculine form *ḥămôr* is used with the feminine form of the verb. Is it possible, then, that in other places where the word *ḥămôr* is used the intention is to the female of the species? Were the she-asses preferred because they could provide milk? She-asses were a sign of wealth, since Job's property includes a large number of them and no he-asses (Job 1:3; 42:12).

23. In Medinet Habu, Rameses III records that, in his wars against the Libyans, he captured asses among many different types of animals (Edgerton and Wilson 1936:67–8). It is quite possible that Libya was a source of donkeys for the Egyptians, rather than Canaan.

24. The exact date of the domestication of the horse is still being debated (Clutton-Brock 1992:55–56). Simpson dates the domestication of the horse to not earlier than 2500 B.C.E. (Simpson 1951:25). Anthony agrees with Clutton-Brock concerning the date of domestication (4200–4000 B.C.E.) and with Clutton-Brock (1981:84) and Simpson concerning the reason, i.e. meat. Like Clutton-Brock, he identifies the region where domestication occurred as the Ukrainian steppes, north of the Black Sea (Anthony 1984, 1991).

25. Like Anthony, Gilbert suggests that the horse was domesticated in South Russia in the late fifth millennium B.C.E., and was brought into the Near East in the early third millennium (Gilbert 1991:104).

26. There is disagreement about the place of origin of the Arabian horse, known also as *kohl* or *kehailan*. The possibilities are debated by Lydekker (1912:154-64) and Simpson (1951:34-41).

27. The employment of cavalry and chariotry in the military produced new defense initiatives. The biography of Amen-em-heb, a soldier under Thusmoses III, recounts that in a battle near Kadesh, "the Prince of Kadesh sent out a mare, which [was swift] on her feet and which entered among the army." He killed her with a dagger. While the footnote explains that the mare was "to stampede the stallions of the Egyptian chariotry" (Pritchard, 1969a:241), it seems more likely that the mare was in heat and was sent to disrupt the battle.

28. When Adonija wanted to become king, he employed chariots and cavalry (1 Kgs 1:5) because this was sign of royalty.

29. Elat believes that David used chariotry, but he hamstrung most of the horses he captured because he could not incorporate them into his forces (Elat 1977:214–5).

30. The terms used for chariot, chariotry, cavalry, steed, horseman are interchangeable and confusing and, at times, can be distinguished from each other only from context.

31. Chariotry for sport was introduced later. At Olympia, as early as 648 B.C.E., there were chariot races, horseback races, and mule-cart races (Anthony, 1984:27; Simpson, 1951:30).

32. Syria as a source for horses was known even earlier. A painting in the tomb of Huy shows a horse among tribute brought by Syrians to Tut-ankh-Amen (Pritchard, 1969b:fig. 52).

33. While these numbers might be considered exaggerated, the same report mentions other forces with much smaller numbers, a fact which suggests that the high numbers might be not too far from the truth.

34. Lachish is mentioned as a chariot city in connection with the murder of King Amaziah (2 Kgs 14:19–20; 2 Chr 25:27–28).

35. This should not be misunderstood as suggesting the Assyrians abandoned chariotry, because they, too, continued to develop the art of chariotry in battle and hunting. Assyrian chariotry employed two soldiers per chariot, one as the driver and the other as a fighting archer (Elat 1977:pl. 8). On Assyrian chariotry see Wiseman 1989:42–3.

36. Assyrian annals and Egyptian campaign reports include booty and tribute lists enumerating horses and chariots captured and transported to the homeland (e.g. Pritchard, 1969b:246–7, 275–8). However, the horses imported by the Assyrians from the east were untrained and were brought in herds, while those acquired in the west were already trained and were placed immediately into military service (Elat 1977:81–2).

37. Elat is not completely positive that the source of these horses was Nubia or Egypt (1977:76, 200–1).

38. Previously, horses were ridden in pairs, one horseman controlled the horses and shielded the other, who was using a bow and arrow (Elat 1977:79 and pl. 6).

39. A relief from the north palace at Kuyunjik shows mounted Assyrians chasing after and shooting at Arabs fleeing on camels (Pritchard, 1969b:fig. 63). A seventh-century B.C.E. grave at Hallstatt (Germany) yielded a scabbard depicting worriors on horseback (Clutton-Brock, 1992:59).

40. See for example the ninth-century B.C.E. relief from Tell Halaf (Pritchard, 1969b:fig. 164). The stirrup was introduced only in the Late Roman-Byzantine period (Anthony 1984:27; Wiseman 1989:44) or the early Middle Ages (Clutton-Brock, 1981:87–88; 1992:93).

41. A bronze bit from Tell el-'Ajjul is dated to the "Early Hyksos Age" (Pritchard 1969b:265, fig. 139).

42. Von Soden suggests that "white horses were considered particularly valuable and were frequently given as gifts to the temples" (1994:93). However, his mention of "the horses that the kings of Judah had set up in honour of the sun at the entrance to the house of the Lord" (2 Kgs 23:11) are not a good example of such a practice because it seems that the reference here is to statues (or figurines), but not live horses.

43. In biblical terminology, a stable is *'urwâh** (1 Kgs 5:6; 2 Chr 32:28) or *'uryâh** (2 Chr 9:25), and a manger is *'ebûs* (Isa 1:3).

44. The two main theories regarding the function of this type of structure claim it was either the royal stables or public storage buildings.

45. See Pritchard's re-assesment of the stable theory (Pritchard 1970; Yadin 1975) and Holladay's detailed response, especially the parts related to the question of horse sizes (1986:122–3, 130 n. 36, 131, 136, 147, 149).

46. For the present-day state of Przewalski's horse, its history, and its distribution. see Mohr 1971.

47. For a discussion of the use of small horses for drawing chariots, see Clutton-Brock 1981:87–88.

48. The existence of horses of different sizes is well-known from Assyrian documents referring to "large horses" imported from different sources (Elat 1977:76-7).

49. For exceptions to this rule see Clutton-Brock 1981:94. Lack of reproductive ability priced mules in the third millennium B.C.E. higher than other equids. While naturally-reproductive equids were priced from four to seven silver shekels, the price of a mule ranged between twenty to thirty shekels (Zarins 1986:185). In Ebla, the highest price of a mule was five mina or 300 shekels (Ibid. 187). Remains of *E. caballus* x *E. asinus* in South Turkey and North Syria indicate "that in the Bronze Age in North Syria mules and/or hinnies were deliberately bred on a large scale to meet the needs of developing trade" (Buitenhuis 1991).

50. The Sumerian and Babylonian terminology related to equids is quite clear: anše = *E. asinus*, anše-DUN.GI or anše-LIBIR = *E. asinus*, anše-eden-na = *E. hemionus*, anše-BARxAN = *E. asinus* x *E. Hemionus*, anše-zi-zi or anše-kura-ra = *E. caballus* (Clutton-Brock, 1992:89). On Sumerian equid terminology, see also Postgate 1986.

51. Mules are mentioned several times in the *Iliad* and *Odyssey* suggesting the possibility of their existence in Greece and Asia Minor during the Trojan war, or at least in the time of Homer. For information concerning modern-day mules, see Davis 1984.

52. It is possible that the mule was considered a royal animal because it was rare and expensive. Furthermore, the production of mules is easier and more successful than that of hinnies (Clutton-Brock, 1981:95).

53. According to Ezekiel 27:14, Tyre imported mules from Togarmah.

54. Textual evidence shows that, during the Ur III period, mules (anše.BAR.AN) were associated with threshing activities (Zarins 1986:183 n. 14). anše.BAR.AN is also used in Ebla (Ibid. n. 15).

55. For the dromedary's ability to withstand hot climate, see Yagil 1993. For a brief background of the dromedary and bactrian camels, see Uerpmann 1987:48.

56. How difficult it was and how long it took to introduce the camel into certain regions can be illustrated by the fact that camels arrived in Egypt only at the beginning of the Persian period (late sixth century B.C.E.); until then donkeys performed all the functions later assigned to camels (Carrington 1972:70).

57. The period of lactation lasts for two years and the daily yield is about two liters, although at the beginning it might reach ten liters (Bodenheimer 1935:125). Köhler reports that lactation lasts twelve months and longer; because camels do not reproduce easily, she maintains that meat was not the primary reason for camel domestication (1984:202).

58. A jar filled with camel dung and fragments of camel's hair fabric, dated to 2000–1600 B.C.E., was discovered at Shahr-i Sokhta in Iran (Bulliet 1975:149, 152). Clutton-Brock dates this find to 2600 B.C.E., insisting that it came from a domestic camel (1981:126). Uerpmann is convinced that since this site yielded a few camel bones, camel dung, and some camel hair used with sheep wool in a piece of textile, the animals were domesticated (1987:55). Köhler suggests that this find illustrates the original reason for camel domestication (Köhler 1984:202, 204). In Egypt, camel hair was used for weaving fabrics, tent-cloth, and sack-cloth at least since the Roman period (Forbes 1956:5, 63).

59. For a discussion of the raḥil, see "herding."

60. A female camel is known in the Bible as bikrâh (Jer 2:23).

61. This use is evident in the stories about Rebecca's journey to Canaan (Gen 24) and Rachel's departure to Canaan when she took "the household gods and put them in the camel-bag and was sitting upon them" (Gen 31:34).

62. The camel's hair is plucked, not shorn, in the spring when it becomes loose. One camel can yield between two to three kgs of hair (Bodenheimer 1935:125).

63. The prohibition does not include drinking camel's milk (or that of any other unclean animal).

64. In preindustrial times and even in the modern period, the camel has been observed being utilized for plowing and other agriculturally related tasks.

65. In addition to camels, "a river ox" (in Akkadian alap nâri, hippopotamus or water buffalo) was also brought as tribute (Pritchard 1969b:fig. 281). Wapnish suggests that it was possible for these Bactrian camels to have come from Egypt, although this was not their known natural habitat (1984:180–1).

66. These same details are in the British Museum Fragment K 8523 (Pritchard 1969b:fig. 292).

67. There is no question that the camels encountered by the Assyrians in Arabia as part of the opposing military or taken as booty and tribute originated there. Evidence provided by Ashurbanipal shows that camels were raised in Arabia: "Even when the camel foals, the donkey foals, calves or lambs were suckling many times on the mother animals, they could not fill their stomachs with milk" (Pritchard 1969a:300).

68. Camel bones from other periods include five bones from the Late Bronze Age, two from the early Iron Age, eight from the Iron Age II (800–700 B.C.E.), eighty-five from the Hellenistic period. No camel bones were recovered from the Chalcolithic and Middle Bronze Age II strata (Wapnish 1984:172–4).

69. Another suggestion is that these camel bones may derive from lame and other unwanted camels sold off by desert traders who used them as pack animals along the spice routes (Davis 1987:166). In an earlier publication, Wapnish compares the evidence with four possible models for the use of camels at Tell Jemmeh and reaches the conclusion that, in the Assyrian period, they were used for transport in the military and later utilized as food and, in the Persian period, they were also used for transport of incense from Arabia (1984:175–9).

70. "Ox" refers here to a castrated bull.

71. Instead of *ṣemed bāqār* (1 Sam 11:7; 1 Kgs 19:21).

72. One way to measure a man's wealth was by the number of teams of oxen he owned (Job 1:3; 42:12).

73. Sometimes translating the Hebrew *pārâh* (cow) into the English 'heifer' gives the impression that the text refers to a young cow, but actually the term is used for a full-grown cow. See also Deut 21:3.

74. The working cow portrayed in these accounts is most likely the so-called Arab cow; see "Large Cattle: Milk Cows" above.

75. Zooarchaeologists suggest the when the bone sample contains at least 20% large cattle remains, it is a sign of an agricultural economy.

Chapter 4

Other Domestic Animals

HOUSE AND YARD ANIMALS

Dogs

The dog (*Canis familiaris; keleb*, Exod 11:7), "man's best friend," was the first animal to be domesticated, about 12,000 (Clutton-Brock 1981:34, 42; Olsen 1985:72–3) or 10,000 years ago (Davis 1987:127).[1] Since then, for a variety of reasons and needs, many breeds have been developed (see, for example, discussion in Wapnish and Hesse 1993:65). Today, there are more than four hundred breeds of domestic dogs (Clutton-Brock 1981:34).

The dog (Fig. 4.1) belongs to the Order Carnivora, Family Canidae.[2] All dogs, whether they are Great Danes or Chihuahuas, are descendants of wolves (*Canis lupus*) that were tamed in different regions by humans some time at the end of the last Ice Age (Clutton-Brock 1981:34, 42).[3] The dog is the only member of this family that barks, and this attribute was utilized in many of the chores assigned to dogs. It seems that barking is the result of domestication.

The association between man and dog led to a special relationship that was probably first connected with hunting activities and later, with the domestication of large and small cattle, was extended to herding. Dogs were highly regarded in the early days of Egypt, as seen from their mummification and interment (Carrington 1972:78; Wapnish and Hesse 1993).

Figure 4.1. *Dogs. (From Bodenheimer 1935:pl. XIII.)*
a. Saluki greyhound. b. Tasi greyhound.
c. Pariah dog, heavy type. d. Pariah dog, light type.

Dogs in the Ancient Near East

While different breeds were developed and different functions were assigned
to them, the dog's first role as man's hunting companion continued throughout history. Tut-ankh-Amen is depicted as a hunter on his chariot, chasing
gazelles and shooting them with arrows while two dogs attack the wounded
animals (Pritchard 1958: fig. 41). The dog's role as a hunting animal was not
exclusive to Egypt. In Mycenean culture (Middle and Late Bronze Age), dogs
were also used in hunting. A fresco discovered in Tiryns depicts greyhounds
attacking a boar previously wounded with a spear (Davidson 1962:292–93).
Assyrian reliefs from Ashurbanipal's palace in Nineveh show that hunting

dogs were restrained with a leash by handlers armed with spears. At the right time, the dogs were released to chase and catch the hunted animals, some of which had already been hit by spears and arrows. The large size of the dogs and their stance portray them as vicious (Hall 1928:pl. liii).

The biblical text gives us some hints about the dog and its place in society. Packs of unmanaged dogs (Ps 22:17) roamed the cities (1 Kgs 14:11; 16:4). Other dogs were used in Israelite society in herding (Isa 56:11; Job 30:1)[4] and as watchdogs (Isa 56:10). However, in spite of their positive contribution, they were not well-treated, as can be seen from Goliath's remark to David: "Am I a dog that you come out against me with sticks?" (1 Sam 17:43). In the biblical context, many times the dog has bad connotations (Exod 22:30; Deut 23:19; 2 Sam 3:8; 2 Kgs 8:13; Prov 26:11; Eccl 9:4), especially a dead dog (1 Sam 24:15; 2 Sam 9:8; 16:9). Nevertheless, sometimes individuals mistreating a dog did not meet with the community's approval (Prov 26:17).

Unmanaged dogs (pariah) were common in the Near East (Wapnish and Hesse 1993:74), and their presence in the urban setting, which is evident in the Bible (Ps 59:7, 15), might have caused ill feelings toward them.[5] It is very likely that the dogs mentioned as eating bodies of the dead and licking their blood (1 Kgs 14:11; 16:4; 21:19, 23–24; 22:38; 2 Kgs 9:10, 36) were pariah dogs. Dogs devouring humans are seen in a wall painting showing Pharaoh Tut-ankh-Amen fighting Nubian enemies from his chariot while two dogs are engaged in a bloody attack on two enemy soldiers (Davidson 1962:112–13).[6]

Extra-biblical texts from the Iron Age add very little to the knowledge of the status of dogs. Watchdogs had collars and were tied, as can be assumed from Ashurbanipal's report of his campaign against Uate', king of Arabia: "Upon the (oracle-) command of the great gods, my lords, I put a dog's collar on him and made him watch the bar (of the city's gate)" (Pritchard 1969a:298). In another report, he says: "I . . . placed a dog collar around his neck and made him guard the bar of the east gate of Nineveh which is called Nîrib-masnaq-adnate" (Pritchard 1969a:300). This must have been a degrading experience.

Watchdogs are also known from Egypt, where they guarded not only against enemies but also protected their owner from other dogs. In a letter, an officer stationed on the border with Palestine during the Eighteenth Dynasty states, "There are 200 large dogs here, and 300 wolfhounds, in all 500, which stand ready every day at the door of the house whenever I go out. . . . However, have I not the little wolfhound of Nahereh, a royal scribe, here in the house? And he delivereth me from them. At every hour, whensoever I sally forth, he is with me as a guide upon the road . . . " (Phillips in Carrington 1972:78).

Dogs were not just used in hunting, herding, or as watchdogs, but were kept also as pets (Fig. 4.2). This can be surmised from the way they are depicted in bronze and ivory figurines and from the names inscribed on the collars they wore (Carrington 1972:78). "The dog of Sumu-Ilum" must have

Figure 4.2. *Pet dog, detail of a scene from the Tomb of Nebamun; Thebes, Egypt.*
(The Oriental Institute, University of Chicago.)

Figure 4.3. *Dog of Sumu-Ilum.* (Musée du Louvre/Antiquités Orientales.)

been one of them (Fig. 4.3). Clay figurines from Mesopotamia showing children riding dogs (Van Buren 1930:figs. 209–10) add support to the notion that some of these animals were kept as pets.

Breeds of Dogs in the Ancient Near East

Since a variety of dog breeds had already been developed even before the advent of agriculture (Clutton-Brock 1981:43), it is not surprising that skeletal remains and artistic representations (some of which have already been mentioned) reveal that several breeds were extant in the Ancient Near East. Only very few dog remains have been discovered in Iron Age contexts, including three skulls at Lachish, one of which might have belonged to a saluki-type dog (Bates 1953:410). Other Iron Age dog remains were found at 'Izbet Sartah (Hellwing and Adjeman 1986), Tel Michal (Hellwing and Feig

1988), Beersheba and Tel Masos (Hellwing 1984:110). Remains of nine adult and three immature canids were found in Ashdod Stratum 3 (Haas 1971). The large number of caprid bones found in the same strata suggests that at least some of those dogs were used in herding.

Dog remains from periods earlier and later than the Iron Age are well attested in the archaeological record. Dog (or jackal) skull fragments were found in two tombs at Megiddo, one dating to the Chalcolithic and the other to Chalcolithic/EB (Bates 1938:210, 212). Other remains from the Chalcolithic and EB I periods were discovered at Gilat (Levy 1991) and at Tell Halif (Seger 1991).

Persian and Hellenistic strata yielded a large number of dog remains; most notable are the burials at Ashkelon, where "the Ashkelon dogs appear to be a variable group similar to most other ancient Near Eastern dogs and modern dogs of no particular ancestry. They do differ from those animals identified as salukis and greyhounds" (Wapnish and Hesse 1993:65).[7]

In Egypt, there were several breeds of dogs, as can be seen from their depiction in various Dynasty XII tombs at Beni Hasan (Wapnish and Hesse 1993:73). One of the earliest depictions of dogs, possibly saluki hounds used in hunting, was found in Wadi Abu Wasil in Egypt and dates to around 6000 B.C.E.. Dogs are frequently depicted in Egyptian art, especially in tombs, and these artistic representations illustrate some of the functions dogs fulfilled. Dogs participating in hunting are depicted in a scene of a dog attacking a gazelle and an antelope that appears on a relief in the mastaba of Mereruka (Dynasty V) at Saqqarah (Michalowski 1968:fig. 247). Two dogs, each attacking a gazelle, appear on a steatite disk, possibly a gaming piece, from a First Dynasty tomb (Davidson 1962:46). All these dogs have a slim body and might have been salukis. In the tomb of Sirenput (Dynasty XII) at Aswan, two dogs of different breeds appear on a stone relief (Michalowski 1968:fig. 308). One dog is taller and has a slim body, like a saluki or greyhound; the second, smaller dog has six teat-like protrusions on its belly and possibly represents a pregnant or nursing female.

Different breeds of dogs were also well represented in Mesopotamia. Several cylinder seals from Mesopotamia show small animals being watched over by shepherds and their dogs (Aynard 1972:52). Mesopotamian hunt scenes depict mastiff dogs chasing wild equids (Fig. 4.4), while being restrained by leashes held by servants (Aynard 1972:49, especially fig. 25; see

Figure 4.4. *Wild asses hunted with mastiffs, from Kyunjik.* (Frankfort 1954:112.)

Figure 4.5. *Mastiff and puppies, Iraq.* (The Oriental Institute, University of Chicago.)

also Clutton-Brock 1981:45) The mastiff breed (Fig. 4.5) is characteristically very large and powerful, deep-chested with a smooth coat.[8] Another example of this breed is that of a steatite statuette of "The dog of Sumu-Ilum" King of Larsa (end of the third millennium B.C.E.) (Fig. 4.3). It has a thick snout and drooping ears, lying down but ready to leap (Aynard 1972:56 and fig. 28). Other breeds of dogs are known in the form of terra-cotta plaques or figurines from Mesopotamia[9] and Elam (Aynard 1972:56–7). A terra-cotta plaque from the beginning of the second millennium B.C.E. shows a child riding an Alsatian-type dog having a sharp muzzle, a thick mane, pointed ears, and a bushy tale. Large sheep dogs with flocks can be seen in outdoor scenes, and there are amulets in the form of dogs with short legs and long or short tails (Aynard 1972:56–7).

Pigs

The domesticated pig (*Sus scrofa; ḥăzîr,* Lev 11:7), a descendant of the wild boar, is a member of the Order Artiodactyla, Family Suidae. Two subspecies of the wild pig were identified for the Middle East, *Sus scrofa attila* Thomas, 1912 and *Sus scrofa libycus* Gray, 1868, the latter ranging south from Anatolia and Syria as far as the Nile Delta.[10] The pig is probably the only animal that was domesticated for consumption only. The earliest remains of the modern pig were found in Middle Acheulean levels at Benot Ya'akov (Hesse 1990:201). Remains of possible domesticated pig are known from Pre-Pottery Neolithic levels in Jericho (ca. 7000 B.C.E.), from Jarmo in Iraq and Argissa-Magula in Greece, where it was found together with remains of early domesticated sheep and goats (Clutton-Brock 1981:72). Remains of domesticated pig were found at several Chalcolithic, Early, Middle, and Late Bronze Age sites in Israel.

Even before the Roman period, pigs were raised in two ways, as herds roaming the forests (Ps 80:14) under the watchful eye of a swineherd and as individuals in a house or sty. "Successful swineherding is associated with rainfall or moist ground, the presence of mixed deciduous forest, and the availability of reserve fodder collected from agricultural activities" (Hesse 1990:204). The pig's two main activities are eating and sleeping, but unlike other animals, the pig can feed continuously for hours and then sleep for many hours. This mode of behavior is helpful in keeping the pig on a schedule similar to that of humans. Another helpful attribute is that pigs will scavenge and feed on foods that dogs and humans can live on (Clutton-Brock 1981:73–4). This makes the pig an easy animal to raise and an efficient meat producer.

In Egypt, as early as the Third Dynasty, pigs were consumed only on certain days and under special circumstances. A somewhat late depiction of pig breeding is a statue from Egypt of a sow and four nursing piglets, dated to ca. 600 B.C.E. (Carrington 1972: fig. 34). In Mesopotamia, there was no known prohibition against eating pork, and Sumerian sites dated to the beginning of the third millennium B.C.E., such as Tell Asmar and Abu Salabikh, yielded remains of domesticated pigs, which were considered primary source of meat (Clutton-Brock 1981:77).

Aynard (1972:50) maintains that wild boars were represented in Mesopotamian art since the earliest times. A glazed green faience from Nuzi of a very realistic boar's head in the round is attached to a disk that might have been the head of an ornamental nail. The figurine, dated to Early Dynastic times, shows signs of having been made in a mold, suggesting the manufacture of many other figurines (Frankfort 1954:143, pl. 139).[11] Another example is that of a small bronze boar, possibly a jewelry piece or clothing decoration, from Iran and dated to probably the sixth century B.C.E. (Kozloff 1981:47–8, fig. 36).

Swine herding is also depicted in Mesopotamian art. One of the earliest examples appears on a seal impression from Mesopotamia, dated to ca. 3000 B.C.E. (Rosen 1995:27), that shows two pigs walking toward two plants (or trees) being followed by dogs, two of which are on leashes held by a man. This scene no doubt depicts swine herding and reinforces the zooarchaeological record. A terra-cotta plaque from the Third Dynasty of Ur period (ca. 2000 B.C.E.) showing a sow suckling her piglets is a good example of pig breeding for meat in that region. It enhances the written record reporting the receipt by the royal palace of "one young sow from the cane-brake for roasting . . . " (Aynard 1972:53).

Extent of and Reasons for Pig Prohibitions

In biblical tradition, the pig is singled out as an unclean animal, and a special prohibition against eating its meat and touching its carcass appears in

Deuteronomy 14:8, "and the pig, because it has cloven hoofs but does not chew the cud, it is unclean for you. You are not to eat their flesh or even touch their dead carcasses" (see also Lev 11:7). The date of this prohibition is hard to determine, but the negative attitude also expressed in Isaiah 65:4 and 66:3, 17 suggests that, at least in the post-exilic Jewish community, the pig was considered unclean. This attitude underlies Proverbs 11:22, "Like a gold ring in a pig's snout is a beautiful woman without good sense." Nevertheless, the presence or absence of pig bones cannot be used to identify a site as Israelite or non-Israelite.

Hesse and Wapnish (1994) suggest that it is possible to identify the ethnicity of a site, "but not on a straightforward presence/absence basis. . . . It is not enough to show that people didn't consume pork and disdained pig, as part of their lifestyle. It has to be demonstrated how those acts were integrated into the social life of the actors as they engaged the larger community in which they lived." Therefore, when dealing with the issue of pig raising, several questions have to be asked:

- Which ethnic communities were open to the consumption of pork?
- How old is the biblical prohibition?
- What were the causes for the prohibition?
- How well was this prohibition observed?
- Can this trend be detected in the archaeological record?

Hesse (1990) and Hesse and Wapnish (1994) suggest several reasons for the prohibition against pig consumption, among them are:

- The political and/or cultural need to preserve a group identity.
- A religious response to cultic practices of another group.
- Environmental and social conditions that favor pig raising in a certain community and elicit the previous two reasons.
- Health considerations, especially fear of trichinosis.
- Economic and political conditions that encourage or discourage pig raising.

Which individual reason, or combination of reasons, was behind the biblical prohibition is hard to determine.[12]

There is no question that pig remains are rare in Iron Age sites in the hill country and the Negev. Some were found in Iron Age Hesban (Jordan) (Uerpmann 1987:table 6a). They are found in large quantities, however, in sites identified as occupied by Philistines, such as Ashkelon, Miqne, and Batash. Nevertheless, the samples show that there is a decline in the number of pigs from the Iron Age I to the Iron Age II (Hesse 1990:table 3). Examining the archaeological record, Hesse observes that in Iron Age Palestine, "pigs are only common in the southern coastal plain, the homeland of the Philistines. There they appear almost suddenly and are associated with non-ceremonial architecture and deposits. Later in the Iron Age the use of pigs declines. . . . Since the increase in pig coincides with the appearance of the distinctive ceramics of the Sea Peoples/Philistines and because the use of pork may be a major foodway of only some sectors of the communities, it is tempting to link the two causally" (1990:219–20).

In addition, Hesse notes that data from the Bronze Age data suggest that restrictions on pig husbandry may have been encouraged by political centralization, something that could have been reused as the Israelite polity began to take shape in the central highlands. Alternatively, pig hate can be related to an important food of the traditional enemy, the Philistines, although this argument is weakened by the lack of evidence that the Philistines were symbolically tied to the animal (1990:219–20).

While Hesse looks at the evidence horizontally, Zeder examines it vertically and deals with the pig issue via analysis of the zooarchaeological record of Tell Halif throughout its whole history, from the Chalcolithic to the present (1996). The ups and downs of pig herding and consumption in this agriculturally marginal area indicate to her that the pig's vicissitudes are related to the political system extant at the site and in the region. Because the pig affords households dietary independence, it appears that its use would be more common when the political system was more loose and open, as in the Early Bronze Age I, and the Late Bronze Age, especially Late Bronze Age IA, "and would decrease in popularity when regional economy was more highly organized and controlled" as in the Iron Age under the Israelite Monarchy (Zeder 1996:305).

Hesse's conclusion concerning the prohibition is that "the suddenness of the appearance of pig use at the three sites [Ashkelon, Miqne, Batash] and its correlation with the settlement of the Philistines support the ethnic theory of pig use. . . . The important point to note is that the use of pig in the Iron Age appears to have been a very restricted event, both temporally and spatially" (Hesse 1990:218). Zeder's conclusion is that "although the decline in pig popularity coincides with the Israelite period, the earlier fluctuations in pig use suggest that Jewish dietary proscriptions cannot account for this pattern" (1996:305). Furthermore, "the reason pig use declines is not due to the degree of agricultural intensification, but rather to the degree of integration into regional economy that most often accompanies periods of agricultural intensification in the Near East" (Zeder 1996:306).

According to Zeder, even when pig raising was frowned upon by the political/social/economic system, there were individuals who would raise pigs on a small scale (1996:306). Therefore, it is not surprising that a skeleton of a domesticated pig was discovered in Hazor south of the Citadel in Stratum VA, dated to the second half of the eighth century B.C.E. (Angress 1960). The skeletal evidence strongly suggests that the better parts of the animal were removed and consumed.[13]

Cats

Cats (*Felis silvestris*) belong to the Order Carnivora, Family Felidae (Fig. 4.6). Most domestic cats are descendants of two subspecies, the European wild cat, *F. s. silvestris*, and the African wild cat, *F. s. libyca* (Clutton-Brock 1981:108). However, it is not known when exactly cats were domesticated (Clutton-Brock 1981:107). One of the problems for identification is that there is hardly any osteological difference between the wild and domestic cats. This is probably because man did not control its breeding as with other animals. Remains of cats from Pre-pottery Neolithic Jericho (ca. 7000 B.C.E.) found together with remains of other animals in association with man suggest some kind of a relationship (Clutton-Brock 1981:111). Other prehistoric sites yielded cat remains in association with other animals, but it is hard to determined whether they were pets or used as food or for their pelts.[14]

Cats were not represented in Mesopotamian art (Aynard 1972:56),[15] but other felines like the leopard and the cheetah were depicted in hunt scenes (Bodenheimer 1960:44, 100).[16] In Egypt, although the domesticated cat was very popular, there is no evidence for its domestication before the Eleventh Dynasty, and not until the New Kingdom that overwhelming numbers of art

Figure 4.6. *Cat.* (From Bodenheimer 1935:pl. xiii.)

objects prove its popularity. One such example is a painting from Thebes (ca. 1420–1411 b.c.e.) of a tabby cat[17] fowling in the marshes (Clutton-Brock 1981: fig. 10.6). It has been suggested that the cat was originally brought into the Egyptian house as a pest-destroyer (Carrington 1972:78–9).[18] The Egyptian cat is reminiscent of its cousin the modern Siamese cat, and is descendent of the African "yellow cat" or "sand cat", F. s. *libyca.* (Bodenheimer 1960:44; Carrington 1972:78 and pl. 7).

The Egyptian cat (Fig. 4.7) was the most sacred animal, and many cats were mummified. The following episode from the nineteenth century might illustrate this. In the Natural History section of the British Museum, there is a cat skull that is the sole survivor of a single consignment of nineteen tons of such cat mummies that was shipped from Egypt to England to be ground for fertilizer and spread on the fields (Clutton-Brock 1981:110 and fig. 10.5).[19]

The cat is known in the Mishnah but is not mentioned in the Hebrew Scriptures.[20] Nevertheless, the proximity to Egypt and Mesopotamia, and the existence of wild varieties suggest that this animal was known in Palestine (Bodenheimer 1935:37).[21] But how common and popular the cat was in Eretz Yisrael is very hard to determine. It seems that cats were extant in regions close to Egypt. This is suggested by remains of six small felids found in Area D, Stratum 3 (Iron Age ii) in Ashdod (Haas 1971:212). A cat mandible was found among human skeletal remains at Vered Jericho, but it belongs to an immature animal and it is hard to tell whether it belonged to a domestic or wild cat (J. Zias, personal communication).

Figure 4.7. *Cat under a chair, detail from the tomb of a 'Harbor-master in the Southern City,' Egypt.* (The Oriental Institute, University of Chicago.)

Cats, like pariah dogs, carry rabies, and as such have not been welcomed by humans in certain cultures. On the other hand, their hunting talents were appreciated when used against mice and rats, but not birds (Bodenheimer 1935:130).

~

The dog and the pig were probably the first animals to be domesticated, and it is very possible that the similar diet of the dog, pig, and humans created the closeness that led to domestication. As for the cat's domestication, very little is known. As herd animals were domesticated, the dog earned its keep by becaming a helper. The dog's association with different cultural groups in different regions led to the development of breeds.

The pig, which was probably domesticated for its meat, never developed another function other than being a meat producer, even though several cultures frowned upon the use of its meat. Although there are several suggestions for the reason behind this prohibition, the real answer still alludes us.

While the domestic cat was very common in Egypt, and to a certain degree in Mesopotamia, very little is known about its place in Eretz Yisrael. Unlike other animals, very few skeletal remains of cats are known there, and the close similarity between the domestic and wild cat prohibits distinction between the bone remains of the two. Conclusions rely very heavily on the context in which the bones are found.

Notes

1. The exact date of the domestication of the dog is hard to determine because of the close resemblance between the early domestic dog and local wild species of the same genus (Olsen 1985:xii). For remains of early domestic dogs (Neolithic) see Olsen 1985:71–8. More recently it has been suggested that the dog was domesticated even earlier than the Natufian period (12,000 B.P.), i.e., the Kebaran (16,000–18,000 B.P. (Dayan 1994a).

2. Other members of the family include the jackal (*Canis aureus*), coyote (*C. latrans*), fox (*Vulpes vulpes*) and different subspecies of the wolf (*C. lupus*). Several of these species are still found in the Middle East in general and in Israel in particular (Dayan, Simberloff et al. 1992). It has been observed that changes through time in the species diversity of carnivores are not due mostly to human intervention but can be attributed to environmental changes such as precipitation, which affects their food sources (Dayan 1993, 1994b).

3. For more on the domestication of the dog, see Clutton-Brock 1981:42–5 and Olsen 1985.

4. The large number of dog bones found in Iron Age strata also containing caprid bones strongly suggests that at least some of the dogs were used as sheep dogs. Such sites include Tel Michal (Hellwing and Feig 1988), Beersheba (Hellwing 1984), and 'Izbet Sarta (Hellwing and Adjeman 1986). Similar relationships can be observed in earlier and later strata such as at Megiddo (Bates 1938), Tel Kinrot (Hellwing 1988-89), and Tell el-Hesi (Bennett and Schwartz 1989).

5. Bodenheimer discusses pariah dogs and cats in preindustrial Palestine (Bodenheimer 1935:128–30). Dogs' susceptibility to rabies might have caused further alienation.

6. This scene is reminiscent of the prophecies concerning the death of Jezebel (1 Kgs 21:23).

7. Other sites with dog remains include Tel Batash, Ben Gurion airport, Dor, Hesban, Hesi, Tel Michal, and Gezer (Bennett and Schwartz 1989; Hellwing and Feig 1988; Wapnish and Hesse 1993:68–9).

8. See also Van Buren 1930:207–8. Representations of mastiffs are known also from Old Kingdom Egypt (Carrington 1972:72).

9. Dog breeding in Mesopotamia is suggested by a terra-cotta plaque showing a female mastiff nursing four puppies (Frankfort, 1954:pl. 59a) and a clay figurine depicting the same scene (Van Buren 1930:fig. 213).

10. Ancient and modern distribution of the pig in the Near East is discussed by Uerpmann 1987:41–45.

11. For more on the wild boar in ancient art and literature, see Bodenheimer 1960:51.

12. In a recent article, Zeder provides a similar list of reasons for the prohibition (Zeder 1996).

13. Rosen notes that pigs were raised during post-Byzantine periods in Palestine and other Mediterranean regions by Christian communities in spite of the prohibition placed by the Moslem rulers on such activity (1995:27–8). Similar behavior might have been prevalent among non-Israelites residing within the Israelite community that abhorred the pig.

14. Cat remains were found in the Lower Levalloiso-Mousterian civilization in the Tabun and Skhul caves (Bodenheimer 1935:21–2 and table 2).

15. According to von Soden, the cat "was only rarely depicted" in Mesopotamian art (von Soden 1994:91).

16. A list of different feline species, including *F. s. lybica*, appears in the Sumerian HAR.RA = *hubullu*, the oldest work on zoology (Bodenheimer 1960:109).

17. There are two types of tabby cats, striped and blotched. The striped (*F. torquanta*) has narrow wavy vertical stripes on the sides of the body, from the shoulders to the root of the tail. The blotched (*F. catus*) has obliquely longitudonal stripes on the sides of the body forming patterns known as "spiral," "horseshoe," or "circular" (Clutton-Brock 1981:108 and figs 10.1, 10.2).

18. In Mesopotamia, the cat played the same role (von Soden 1994:91).

19. See also Bodenheimer 1960:44, 127.

20. In Sumerian, it is known as "sa-a," and in Akkadian as "suranu" (von Soden 1994:91).

21. There are four small species of wild cats presently found in Israel: the caracal (*Felis caracal*), the jungle cat (*F. chaus*), the wildcat (*F. silvestris*), and the sand cat (*F. margarita*). Among the large cats, only the leopard (*Panthera pardus*) is still found in Israel. While the lion (*Panthera leo*) became extinct in the Middle Ages, the cheeta (*Acinonyx jubatus*) was last seen about forty years ago (Dayan et al. 1990:41).

Chapter 5

The Birds and the Bees

BIRDS

Throughout history, bird hunting was very popular all over the Near East for food and, to a certain extent, for acquiring singing pets (Bodenheimer 1960:56). Bird hunting was done in several ways, and a metaphor in Psalms 140:5 refers to two different techniques: "The arrogant set hidden traps for me; villains spread their nets and lay snares for me along my path."[1] According to biblical references, most birds eaten in the Iron Age were wild rather than domesticated, but the archaeological record suggests that the ratio was even. The definition of birds[2] permitted for consumption (Fig. 5.1) is very broad (Deut 14:11) and is delineated by enumerating the unclean birds (Fig. 5.2). It is interesting to note that the bat was considered a bird (Fig 5.3).

According to the Bible, all birds could be eaten except for those that were defined as unclean, and these were presented in two lists. The one in Leviticus 11:13–19 includes twenty species, while Deuteronomy 14:12–18 includes twenty-one.[3] In general, birds unclean to eat are birds of prey and those that feed on carrion and fish (Bodenheimer 1960:199). However, it is interesting to note that several of the unclean birds, like the eagle and the raven, play positive roles in biblical metaphors and other symbolic instances.

With the exception of the following, the two lists are similar in names and order. Leviticus 11:14 has *dā'âh* and *'ayyâh,* while Deuteronomy 14:13 has *rā'âh, 'ayyâh* and *dayyâh,* in this order. The *'ayyâh,* which is common to both lists, is identified by Shulow as the kite (*Bueto sp.*) (1967). The remaining three terms are probably a reference to the *dayyâh* (Isa 34:15) which is identified as the falcon, *Milvus migrans* (Shulow 1967; Anbar 1970:182). The eighteen birds common to both lists are identified in table 5.1.[4]

Table 5.1. *Unclean birds, in Hebrew, Latin, and English nomenclature, listed in both Leviticus 11:13–9 and Deutoronomy 14:12–8.*

Hebrew	Latin	English
nešer	*Gyps fulvus*	Griffon vulture[a] (Fig. 5.2f)
peres	*Gypaetus carbarus*	Lammergeier
ʿăzniyyâh	*Aegyptius monachus*	Cinereus vulture
ʿôrēb	*Corvus sp.*	Raven[b]
bat yaʿănàh	*Struthio sp.*	Ostrich[c]
taḥmās	*Falco sp.*	Falcon[d] (Fig. 5.2e)
šāhap	*Strix butleri* or *Larus sp.*	Wood owl or gull
nēṣ	*Accipiter sp.*	Hawk
kôs	*Athenae noctua*	Tawny owl[e] (Fig. 5.2c)
šālāk	*Ketupa ceylonesia*	Fish owl
yanšûp	*Asio flammeus*	Short-eared owl
tinšemet	*Tyto alba*	Barn owl (Fig. 5.2d)
qāʾāt	*Pelicanus sp.*	Pelican (Fig. 5.2g)
rāḥām or *rāḥāmâh*	*Neophron percnopterus*	Egyptian vulture[f] (Fig. 5.2b)
ḥăsîdâh	*Ciconia ciconia*	Stork
ʾănāpâh	*Ardeidea gen.*	Heron[g] (Fig. 5.2h)
dûkîpat	*Upopa epops*	Hoopoe[h] (Fig. 5.2a)
ʿăṭallp	Order *Chiroptera*	Bat[i] (Fig. 5.3)

Sources: Bodenheimer 1935, 1960; Shulow 1967; Anbar 1970.
a. The vulture was the symbol of royalty in Egypt and Mesopotamia, and the Book of Ezekiel (17:3). Vultures were also representative of Mesopotamian gods (Van Buren 1945:80), and they appear on several stelai (Pritchard 1969b:figs 298–9, 301).
b. Ravens and crows were used for divination; see for example the

role of the raven during the flood (Gen 8:6). Notice also the positive role played by the ravens when they fed Elijah in the wilderness (1 Kgs 17:6).

c. Until recently, the ostrich could be found in Transjordan and Arabia. Its eggs were used as food and for making cups and other vessels (Bodenheimer 1960:59). The ostrich was represented on several monuments and can be seen in bas-reliefs as embroideries on robes of Assyrian kings. In addition, in Mesopotamia, ostrich eggs were treasured as art objects and luxury items and were attributed magical powers (Aynard 1972:50).

d. Falconry was apparently a royal sport in Assyria during the seventh century B.C.E. (Bodenheimer 1960:54). In Egyptian art, falcons are well-represented. See, for example, Steindorff 1946:figs. 354–5.

e. An example of an owl can be seen on a limestone plaque from Dendera showing the bird facing right, with its head in front view (Steindorff 1946:fig. 356).

f. The Egyptian vulture is depicted many times, especially with the uraeus, as can be seen in Steindorff 1946:figs 350–3.

g. See the depiction of a seated young heron on a limestone plaque from Upper Egypt in Steindorff 1946:fig. 360.

h. Hoopoe was sacred in Egypt, and its unclean status might be a reaction to this status (Bodenheimer 1960:56).

i. In addition to bats being included as birds, grasshoppers are introduced in the Leviticus list under the term *'ôp*, 'fowl': "All winged creatures that swarm and go on all four are prohibited to you" (Lev 11:20).

Three wild birds are mentioned in narratives and sacrifice lists as allowable for sacrifice, and therefore for consumption:[5] *śĕlāw* (quail; *Coturnix coturnix*, Exod 16:13; Num 11:32), *yônâh* (dove; *Columba* sp., Lev 1:14), and *tôr* (turtle-dove; *Streptopelia turtur*, Lev 5:7).[6] Although the rock partridge (*qôreʾ, Alectoris graeca*), which is related to the quail and is native to and common throughout the Middle East (Anbar 1970:274–80), is not mentioned as fit for sacrifice or the table, it is hard to believe that it was not eaten since it was hunted (1 Sam 26:20) and appears in archaeological contexts (Kolska-Horwitz and Tchernov 1989:147–8).

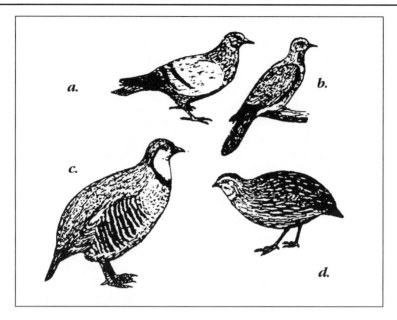

Figure 5.1. *Birds for sacrifice and consumption.* (From Bodenheimer 1935:pl. XVI.)
a. Dove, Columba livia. b. Pigeon, Streptopelia senegalensis.
c. Quail, Alectoris graeca. d. Partridge, Coturnix coturnix.

Bird eggs were also considered food, but biblical references indicate only the collection of eggs of wild birds (Isa 10:14), and one reference (Deut 22:6) allows the removal of the eggs but not of the nesting mother. This must be a sign of environmental considerations against extinction, since the mother is set free to procreate. The collection of bird eggs was quite common in the Near East. Birds' eggs are represented in Egyptian art in images showing the offering of bowls filled with large eggs, possibly ostrich and pelican (Brothwell and Brothwell 1969:54 and fig. 18).

There is no question that, during the Iron Age, some birds were domesticated or were raised under controlled conditions. There are several references in biblical and extra-biblical sources to fattening birds for consumption, probably for special occasion. One reference to such practice is contained in the description of the food prepared daily to be served on Solomon's

Figure 5.2. *Unclean birds.* (From Bodenheimer 1935:pls. xv, xvii.)

a. *Hoopoe, Upopa epops.*

b. *Egyptian vulture, Neophron percnopterus.*

c. *Tawny owl, Athena noctua.*

d. *Barn owl, Tyto alba.*

e. *Falcon, Falco peregrinus pelegrinoides.*

f. *Griffon vulture, Gyps fulvus.*

g. *Pelican, Pelecanus onocrotalus.*

h. *Heron, Ardea pupurea.*

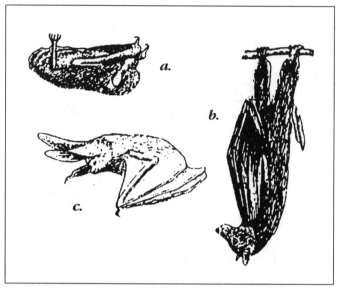

Figure 5.3. *Bats.* (From Bodenheimer 1935:pl. VII.)
   a. Vespertilionid bat.
   b. Fruit bat, Rousettus aegyptiacus.
   c. Long-eared bat, Plecotus auritus.

table (1 Kgs 5:3 [4:23]) which includes *barburîm 'ăbûsîm,* "fattened fowl."[7] Fattening fowl, like fattening other animals, was done by providing special food while keeping them caged or tied to restrict their movements. One allusion to birds being fattened in a cage appears in Jeremiah 5:27–28.[8] In this case, the prophet describes those who take advantage of others saying that "their houses are full of fraud, as a cage is full of birds. They grow great and rich, sleek and bloated." Mesopotamian figurines of geese and ducks, many used as weights, strongly suggest that they were elements of the barnyard. Other Assyrian figurines illustrate sacrificial birds, possibly swans, raised on the farm (Aynard 1972:53). It makes sense that doves, mentioned above as used in Israelite sacrifices, also must have been raised in a domestic environment, because hunting could have damaged the bird and prevented its fitness for sacrifice.

Fowling

Hunting birds is mentioned several times in the Bible, though mostly in the late books. The movements and migrations of wild birds were carefully watched and noted (Jer 8:7; Song 2:12) and birds were hunted (Prov 6:5; Lam 3:52) mostly by traps, *paḥ* (Ps 124:7; Prov 7:23). The eighth-century B.C.E. prophet Amos provides us with a good picture of bird trapping when he says, "Does the bird fall into the trap on the ground (*paḥ hāʾāreṣ*) if no bait (*môqēš*) is set for it? Does a trap spring from the ground and take nothing?" (Amos 3:5).[9] The mention of quail (*šēlāw*) as a source of food in connection with crossing the Sinai Desert (Exod 16:13; Num 11:32; Ps 105:40) reflects the bird's north-to-south migration in the fall. Quail trapping is still being done there today.[10] Not all birds hunted were meant for consumption. Exotic birds such as parrots (*tukkiyyîm*) were brought by expeditions to faraway places (1 Kgs 10:22) as items of luxury and prestige.

The fowling practiced in Ancient Israel must have followed methods similar to those practiced in other parts of the Near East and depicted in ancient art (Fig. 5.4). A scene of bird hunting is known from the tomb of Ka-gemni in Sakkarah (beginning of Sixth Dynasty, 2350–2200 B.C.E.), where a cage-like net filled with birds is spread over a marsh area with the fowlers hiding behind a blind (Pritchard 1969b:271 and fig. 189). Another fowling scene, painted in the tomb of Neb-amun in Thebes (ca. 1400 B.C.E.) shows the activity taking place in a marsh, with papyrus plants jutting out of the water where fish are swimming (Carrington 1972:76 and pl. 5). The possible results of fowling in the marshes can be seen on two New Kingdom relief fragments from Deir el-Bahri, one showing three offering bearers, two of whom each carry a live goose (Steindorff 1946:fig. 243), and the other showing the head of a god in front of which are two live ducks in a net (Steindorff 1946:fig. 271). That the Israelites used nets for fowling is alluded to by the prophet Hosea, who says: "Wherever they turn, I shall cast my net over them and bring them down like birds; I shall take them captive when I hear them gathering" (7:12).

Hunting birds with bow and arrow is seen in a painting on a box from the tomb of Tut-ankh-Amen (ca. 1361–1352 B.C.E.), where he is shown on his chariot chasing a herd of ostriches (and gazelles) (Pritchard 1969b:271 and

Figure 5.4. *Fowling with a trap manipulated behind a blind. Detail from Tomb 3 at Beni Hasan.* (From Newberry 1893:pl. xxxiii.)

fig. 190).[11] Ostriches are very common on cylinder seals and their impressions. One such impression shows an adult ostrich and a young one being chased (possibly hunted) by a heroic figure holding a stick in its raised right arm (Frankfort 1954:fig. 75c).[12] In Mesopotamia, fowling is frequently mentioned in texts. The primary means for fowling were nets and snares, although there also are depictions of people shooting at birds. Sometimes decoys were used to attract birds within range of the hunters; when falcons were employed to capture birds, it was mostly on royal hunts (von Soden 1994:89–90).

Birds were a common motif in ancient Near Eastern art and both real and imaginary birds are depicted. For our purposes, it is important to note the two illustrations of birds from el-Jib, both incised in seventh-century B.C.E. sherds. One shows a small bird inside a six-pointed star ("Star of David"). The particular type of bird is hard to determine, but it could signify a ground or song bird. The other incision is in a cooking-pot handle and shows a bird (Pritchard 1969b:figs 792–93), most likely a chicken (*Gallus domesticus*). A well-defined rooster is depicted on a well-known Iron Age ii seal from Tell en-Nasbeh (Pritchard 1969b:fig. 277; Zorn 1993:1101). The seal, belonging "to Jaazaniah, servant of the king," possibly the officer who reported to Gedaliah at Mizpeh (2 Kgs 25:23), shows a cock in a fighting stance (Fig.

Figure 5.5. *Seal of 'Jaazaniah, servant of the king,' from Tell en-Nasbeh, depicting a fighting cock.* (Pritchard 1969:fig. 277.)

5.5). The latter two illustrations suggest that the chicken was known in Israelite circles, at least, toward the end of the Iron Age II. Another seal, identified as Phoenician and possibly dated to the eighth century B.C.E., depicts two well-formed roosters facing each other ready for fighting (Aufrecht 1995). This also places the domestic chicken in the ancient Levantine barnyard at the second half of the Iron Age II.

When did the domestic chicken appear in this region? Brothwell and Brothwell suggest that the red junglefowl (*Gallus gallus*), which is indigenous to northern India, was the ancestor of domestic fowl (*G. domesticus*) and was known in Egypt by the time of the Old Kingdom (1969:55). Early evidence of chickens in Egypt appears as a drawing of a well-formed rooster on a sherd from the tomb of Tut-ankh-Amen (ca. 1350 B.C.E.) (Brothwell and Brothwell 1969:fig. 19). Von Soden suggests that the chicken was introduced to Mesopotamia from India in the first millennium and is depicted in later periods, but there was no term in Akkadian for the domestic chicken (von Soden 1994:96).[13]

West and Zhou, who studied the time and place of the chicken's domestication suggest that it was first domesticated from the red junglefowl in Southeast Asia some time before the sixth millenium B.C.E. and from there was introduced to China ca. 6000 B.C.E. In India, it was domesticated independently, or arrived there by diffusion, around 2000 B.C.E. (1988). However, evidence from sites in Turkey and Syria indicates that the domestic chicken was present in this region before its domestication in India. As for Palestine, the earliest chicken bones are present in Iron Age I strata in Lachish and Tell Hesban (West and Zhou 1988).[14]

Ancient Ecology and the Role of the Zooarchaeological Record
Evidence of bird bones can help in understanding not only the human diet, but also ecological conditions extant at the time. Bone remains at Tell Jemmeh can be grouped into six clusters: water birds, ground birds, ostriches, gulls, raptors, and chickens. The water-bird group includes ducks, geese, herons, cranes, coots, rails, cormorants, curlews, and corncrakes, a collection that implies the existence of local ponds, streams, and other watery habitats in the region. The ground-bird group includes quail, grouse, partridge, pigeons, doves, and thrushes, a range of species adapted to open brush terrain. During the Middle Bronze and Late Bronze Ages, bones of water birds are seven times as frequent as those of ground birds, while after the beginning of the Iron Age, ground birds are one-and-one-half times more frequent than water birds (Wapnish 1993:428). The changes in the make-up of the remains strongly suggest a change in the local ecology that effected the habitat of the birds and, most likely, life in general.

Fowl remains from the Ophel in Jerusalem suggest that, during the Iron Age, both wild and domestic birds were consumed by the city's inhabitants. Remains from Building D include a high number of domestic birds (78%), while the remains from Building C show a more even distribution of species (48% domestic and 52% wild). The domestic types include domestic fowl (= chicken, *Gallus domesticus*),[15] goose (*Anser* sp., probably *Anser anser*),[16] and duck (*Anas platyrhynchos*); the wild types include chukar or rock partridge (*Alectoris chukar*), pigeon (*Columba livia*),[17] dove (*Streptopelia* sp.), and a passerine (= song) bird (probably *Sylviidae*, warbler family) (Kolska-Horwitz and Tchernov 1989b). The bone evidence shows that the inhabitants of this large city relied for food not only on domestic species, but were also either engaged in fowling or acquired wild birds from fowlers.

RODENTS, REPTILES, AND INSECTS

Man has been eating rodents since early times, and two species were domesticated just for that purpose—the Guinea pig and the dormouse (Clutton-Brock 1981:150–3). It should not be surprising, then, that Biblical instructions prohibit the consumption of "creatures that swarm on the ground . . . : the mole-rat (*ḥoled*), the jerboa (*ʿakbār*),[18] and every kind of thorn-tailed lizard (*ṣāb*); the gecko (*ʿănāqâh*), the sand-gecko (*koăḥ*), the wall-gecko (*lĕṭāʾâh*), the great lizard (*ḥomeṭ*), and the chameleon (*tinšemet*)" (Lev 11:29–30). Additionally, although humans had long been eating different kinds of insects, the Bible declares that "all creatures that swarm on the ground are prohibited; they must not be eaten. All creatures that swarm on the ground, whether they crawl on their bellies or go on all fours or have many legs, you must not eat, because they are prohibited" (Lev 11:41–44). From the biblical point of view, insects in general were considered dirty and defiling.

However, while other insects were prohibited, those of the grasshopper family were allowed to be eaten (Lev 11:20–23) (Fig. 5.6). The locust was considered a delicacy; and, on one Assyrian bas-relief, servants can be seen carrying, among other foodstuffs, long pins of skewered locusts to a royal feast (Aynard 1972:60).[19] Another relief from the palace of Ashurbanipal shows two servants, one of whom is carrying a cluster of pomegranates and the other rows of locusts (Brothwell and Brothwell 1969:fig. 24). Dried

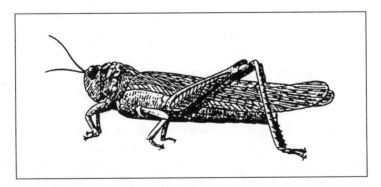

Figure 5.6. *Grasshopper, Anacridium aegyptium (Order Orthopter: Family Acrididae.)*
(From Bodenheimer 1935:pl. xxxv.)

locusts contain "up to 75 per cent protein and about 20 per cent fat; 100 grams of locust, when analyzed, showed the presence of 1.75 mg. of riboflavin and 7.5 mg. of nicotinic acid (vitamin B_2 complex), demonstrating that they are also of value for their vitamins" (Brothwell and Brothwell 1969:68–9).

In the Near East there are two types of locust, the small Moroccan locust (*Dociostaurus maroccanus*), which is the great destroyer of cereal fields and the large desert locust (*Schistocerca gregaria*) which damages fruit trees. The latter is the type used as food (Bodenheimer 1960:77). Biblical references are quite clear about the damage the locust causes (Joel 1:4; 2:25); and in Mesopotamia, an Assyrian relief depicts a nobleman praying for deliverance from the pest or offering thanks for the rescue from a plague of locust (Neufeld 1980:fig. 5). Since, according to the biblical prescription, the locust was edible, it should not be surprising that the insect is depicted on a Judean seal inscribed "belonging to Azaryahu (of) the locust (family)" with an engraved grasshopper under the inscription (Avigad 1966; Neufeld 1978:fig. 1).

While the locust was an edible insect, other insects were useful in different ways. The red dyes kermes (biblical *šānî*, 'scarlet') and cochineal (biblical *karmîl*, 'crimson'; 2 Chr 2:6, 13; 3:14) were produced from the *Kermococcus vermilio Planch.* (biblical *tôla'at šānî,* Exod 26:1), also known as *Kermes nahalali* and *K. greeni* (Bodenheimer 1935:306; Bodenheimer 1960:157), that lived on the kermes oak (*Quercus coccifera* L.) native to the Mediterranean coast and the Near East. The dyes are "extracted from the bodies of female insects belonging to the family Coccoidae" (Forbes 1956:100) and their larvae (Koren 1996). Others maintain that they are made of the unlaid eggs of the insect (Barber 1991:230). These dyes range from a brilliant red to scarlet.

In the Neolithic grotto of Adaouste in France, textile fibers dyed with kermes were found together with food such as barley, meat, and kermes acorns, the latter used most likely for medicine (Forbes 1956:102). In Israelite religion, scarlet thread was used in certain cultic ceremonies of purification (Lev 14:4,6,49,51–52). The scarlet thread must have had some symbolic meaning since, in the story about the destruction of Jericho, Rahab used it to mark her house for protection from the Israelites (Jos 2:18–21). The Assyrians were introduced to the dye during the time of Sargon II, who returned from invading Urartu in 714 B.C.E. claiming to have taken booty including "red stuffs from Ararat and Khurkhi" (Forbes 1956:102).

Bees

Beekeeping (apiculture) is not mentioned in the Bible, but the main product of the bee (*Apis mellifera*), honey (*děbaš*), is mentioned numerous times. However, that term can be used not only for bees' honey, but also for honeydew and its crystallized form (manna), syrup made of dates or other fruit, or anything sweet. When considering mentions of honey in the Bible, "unless the context makes clear a connection with hives, bees, or honeycomb, caution is warranted" (Crane 1975a:453).

While people use plants and animals as food, only honey and milk are complete foods by their nature.[20] Honey contains pure sugar and small amounts of calcium, phosphates, iron, sulfur, and the vitamins C and B (including the B_2 complex). Its most important component, sugar, contains about 34% dextrose, 40.5% laevulose, 1.9% sucrose, and about 1.5% dextrins and gums, thus the three sugars total about 76.4% (Brothwell and Brothwell 1969:73).

The earliest pictorial record of bees and honey appears in Çatal Hüyük in Anatolia (Crane 1975a:453) and is dated to about 7000 B.C.E. (Crane 1983:35).[21] There is evidence from Mesopotamia that the Sumerians in the fourth and third millennium B.C.E. used honey for medical purposes and in rituals, and it is mentioned in the Babylonian Code of Hammurabi (18th century B.C.E.). However, beekeeping was not practiced in Mesopotamia; only one attempt was made, around 760 B.C.E. by the governor Shamash-Res-Usur with no positive results (Bodenheimer 1960:79; Dalley 1984:85, 203, 208; Frame 1995:281).[22] For this reason, it is puzzling why Isaiah, when speaking about the day of the Lord, says, "On that day the Lord will whistle up flies (*zěbûb*) from the distant streams of Egypt and bees (*děbôrâh*) from the land of Assyria" (7:18). It would have been more appropriate to use the bee as a metaphor for Egypt, rather than for Assyria.

There is evidence that beekeeping was practiced in several parts of the Near East. The Hittite Code (14th century B.C.E.) mentions bees and honey and makes many references to beekeeping (Crane 1975a:454–5). As far as the Hittites were concerned, one 'zipittani' of honey cost one shekel (Brothwell and Brothwell 1969:77). Bees were known in Egypt as early as the

First Dynasty, when the bee became the symbol of the king of Lower Egypt.[23]
In Egypt, there are several pictorial representations describing beekeeping
and related activities (Neufeld 1978:225–37, figs. 2, 6–9). Comparing these
artistic representations with what is known from other ancient cultures around
the Mediterranean[24] and from present-day preindustrial societies, it is safe to
say that beekeeping practices have not changed in a major way in the last five
thousand years. Although there is no direct evidence of beekeeping in Canaan
and Ancient Israel, it is safe to assume that this activity took place in a manner
similar to that which was common in the neighboring ancient cultures.[25]

While the honey referred to in the description of Canaan as "a land
flowing with milk and honey" (Exod 3:8, 17)[26] might be a sweet concoction
made of dates (Borowski 1987:127), there are several references to bees'
honey. Two of the better known refer to honey made by wild bees. The first
appears in Judges 14:8–9 and describes honey made by bees swarming in the
body of a lion killed some time earlier by Samson: "He scraped the honey
into his hands and went on, eating as he went." The second occurrence describes an event in which Jonathan, Saul's son, in spite of and without knowing his father's prohibition against eating anything before the end of the battle
against the Philistines, saw a honeycomb (ya'rat haddĕbaš; see also Song
5:1) in the countryside, dripping with honey, "and he stretched out the stick
that was in his hand, dipped the end of it in the honeycomb, put it to his
mouth, and was refreshed" (1 Sam 14:26–27).[27]

The native Egyptian honeybee, Apis mellifera lamarckii, is known as
an "aggressive" bee (Crane 1983:39).[28] Fraser describes the Egyptian bee as
"of medium size, the first three segments of which are light yellow to reddish
in colour, with black border . . . its abdomen is coloured with greyish-white
hairs. The abdomen of the queen is marked with reddish brown on the first
segment, and the colour areas are variable" (1931:6). The closeness of Canaan
to Egypt suggests that the same type of bee was probably kept there. The
description in Deuteronomy 1:44 of "the Amorites living there [who] came
against you and swarmed after you like bees . . . " is most likely in reference
to that type of aggressive bee. The aggressive nature of the Palestinian bee
can be seen also from the description of the enemies surrounding the Psalm-
ist: "They swarmed round me like bees" (Ps 118:12).

Since no object or installation has been found in Israel that can be interpreted as related to beekeeping in the Iron Age, one needs to study what is

known from neighboring cultures and from ethnoarchaeology. Two types of hives were in use, horizontal and upright. They were made of baked and unbaked clay and in most cases the bees were "encouraged" to form the honeycombs across the hive and not along it.[29] A wall painting from the tomb of Rekhmire (ca. 1450 B.C.E.) shows the use of a smoker during the collection of honey (Crane 1983:39, and fig. 16).[30] Smoke drives away the bees and enables the beekeeper to work in and around the hive. Egyptian pictorial representations show that honey was stored in jars. In 1 Kgs 14:3, a closed vessel (*baqbuq děbaš*) is mentioned as a storage utensil for honey. Jars must have been used for the storage of large quantities of honey referred to in 2 Chronicles 31:5 and Jeremiah 41:8. The quality of the honey produced in Canaan must have been excellent, because otherwise it is hard to understand why, according to the biblical account, Jacob would send to Egypt, a center of honey production, presents that included honey (*děbaš neko't*, Gen 43:11).[31] Furthermore, honey must have been produced there because the Egyptians, under Thutmoses III in his seventh campaign in Phoenicia, found that every port town "was supplied with good bread and with various (kinds of) bread, with olive oil, incense, wine, honey, fr[uit] . . . " (Pritchard 1969a:239) and part of the tribute sent to this king from this region included large numbers of jars of honey (Neufeld 1978:225).[32]

Use of Honey and Other By-Products
In addition to eating and spicing drinks, honey was used in rituals and for medical purposes. About 500 out of 900 Egyptian remedies included honey. The same was practiced in Mesopotamia (Crane 1983:42). Wax (*dônag*; Micah 1:4; Ps 22:15; 68:3; 97:5), a by-product of beekeeping, was used in embalming (Crane 1975a:461) and in casting metal and glass objects in the lost wax method.

There has been a suggestion that bees were used in the Iron Age as a biological weapon for offensive and defensive purposes (Neufeld 1980), but this proposition is very unconvincing and lacks any evidence.[33]

∼

The economy of the Ancient Near East, including Israel, exploited not only large animals but also insects. While most insects, excluding grasshoppers, were not edible, several of them were used in other ways. A certain insect, the *Kermococcus vermilio Planch.*, was used in the production of dye, while bees were kept for their production of honey and wax. Both of these products had several uses and merited mention in tribute lists and other records.

Notes

1. Other allusions to birding can be found in Jeremiah 5:26 and Proverbs 6:5.

2. The collective term for fowl is ʿôp (Jer 5:27; Hos 9:11), and the generic term for bird is ṣippôr (Gen 7:14).

3. There might be a misspelling in Deuturonomy 14:13 resulting in the addition of one species.

4. Milgrom follows others who suggest that birds are prohibited because of their eating and living habits. He also provides a list of possible identifications (Milgrom 1991:661–64). Many of the biblical entries include the additional phrase "every kind of . . . " which suggests that the formulators of the dietary rules were aware of the fact that several of these were families having subspecies.

5. Bodenheimer also notes the absence of domestic fowl in sacrifices (1960:209). However, with the recent discovery of large columbaria at Hellenistic Maresha and other sites, the question should be raised of when doves and pigeons were first raised domestically. This date should be related to the date of raising other domestic fowl such as chickens.

6. Another term connected with a sacrificed bird is gôzāl, mentioned once in Genesis 15:9.

7. The term barbur is sometimes translated as capon, goose, swan, guinea-hen, cuckoo, or young chicken. However, some suggest that the term is a sound-imitating noun, referring to a bird that makes much noise. Others suggest that the origin of the term is the root BRR, 'purify, select,' namely specially selected birds.

8. An extra-biblical reference to caging birds is included in Sennacherib's description of his siege on Jerusalem, where he says that he kept King Hezekiah prisoner in his own city "like a bird in a cage" (Pritchard 1969a:288).

9. The word môqēš is very often used in parallelism to paḥ (e.g., Jos 23:13; Isa 8:14; Ps 69:23) or by itself as 'snare' (e.g., Exod 10:7; Deut 7:16; Judg 2:3).

10. A depiction of a quail chick can be seen in a relief from Luxor, in which it represents the hieroglyphic letter W (Steindorff 1946:fig. 358).

11. For a list of birds identified from Egyptian paintings, statues, and bas-reliefs, see Carrington 1972:76–7. Hunting birds with bow and arrow in Assyria appears on a relief from the palace of Sargon II at Khorsabad (Pritchard 1969b:fig. 185).

12. Another scene of an ostrich, and other animals, being hunted by a man armed with a spear is also seen on a cylinder seal (Frankfort 1954:fig. 75b).

13. The term *tarlugallu* may at times be used to refer to the domestic rooster (von Soden 1994:96 n. 16). The word *śekwî* in Job 38:36 is translated by some as 'rooster'.

14. At these sites, chicken bones are present also in Iron Age II strata (West and Zhou 1988). I would like to thank Liora Kolska-Horowitz for bringing this study to my attention.

15. Although the exact domestication date of the chicken is still unclear, Iron Age strata such as at Tel Michal (Hellwing and Feig 1988) and Beer-Sheba (Hellwing 1984) yield limited remains of this bird.

16. One bone of a domestic goose was recovered at the Lachish sanctuary (Lernau 1975).

17. The question of the domestication of the pigeon is mentioned above in note 14.

18. Usually translated 'mouse' (Brown, Driver and Briggs (BDB): 747). The mouse was considered an agricultural pest and a carrier of disease (Borowski 1987:156). Several types of mice and rats are mentioned in the Sumerian HAR.RA = *hubullu*, the oldest work on zoology (Bodenheimer 1960:110).

19. The locust is considered one of the worst agricultural pests in the Near East (Bodenheimer 1960:77; Borowski 1987:153–56).

20. The child Zeus, after being saved from Cronus, his father, who wanted to swallow him, was hidden in a cave and was fed on milk and honey.

21. Neufeld presents pictorial evidence from Spain of honey gathering dated to the Neolithic (and even possibly to the Paleolithic) period (see also Brothwell and Brothwell 1969:fig. 26, and Neufeld 1978:229–30, fig. 5).

22. Almost all scholars agree that this attempt failed, but Neufeld assumes that it succeeded and that the record shows that Shamash-Resh-Usur learned to boil the honey and separate it from the wax of the honeycomb (1978:238–9, fig. 10).

23. Bodenheimer thinks that the symbol is that of a wasp (1960:74). See also Borowski 1983.

24. The earliest preserved records of beekeeping were written by Aristotle (384–322 B.C.E.) (Fraser 1931:13–28).

25. Macalister claims that he found several jars that he can only interpret as beehives. Unfortunately, the example he presents (Macalister 1912:67, fig. 262) does not look like a potential beehive and has no similarities to any beehives known from the Ancient Near East.

26. See also Exod 13:5; 33:3; Lev 20:24; Num 13:27; 14:8; 16:13–14; Deut 6:3; 11:9; 26:9, 15; 27:3; 31:20; and more.

27. Another term for honeycomb is *ṣûp* as in *nopet sûpîm* (honey from a honeycomb, Ps 19:11; Prov 16:24; 24:13; Song 4:11).

28. According to Bodenheimer (1960:79) the bee native to Egypt is *Apis unifasciata fasciata*, which originated in Africa. Nevertheless, he admits that the Syrian bee *A. mellifera* var. *syriaca* is famous for its stinging.

29. Clay hives from ancient Greece show that the inside of the vessel was roughened by incisions or combing before firing, thus encouraging the bees to build the honeycombs accordingly (Jones 1976:80, 82, 84).

30. Smoking bees away from their hives can be seen in a scene from the sun-temple of Ny-woser-Re (Fifth Dynasty, 2600–2400 B.C.E.) (Neufeld 1978:232–33, fig. 6). Another beekeeping scene is known from the tomb of Pa-bu-sa in Thebes (seventh century B.C.E.) (Brothwell and Brothwell 1969:76).

31. This must have been honey spiced with gum tragacanth. It is hard to believe that Jacob would send to pharaoh wild honey since this could have had a toxic effect. There are several documented occurrences, the earliest of which was recorded by Xenophon after his campaign against Persia in 401 B.C.E. According to him, his soldiers ate wild honey and became "like intoxicated madmen." This could be attributed to honey produced by bees that collected nectar from rhododendron blossoms (Mayor 1995:32–3).

32. Nevertheless, large quantities of honey can also be produced by gathering from the wild (Neufeld 1978:228–9).

33. In connection with this theory, see my suggestion concerning the same biblical references (Borowski 1983).

Chapter 6

Water Fauna

FISH

Eretz Yisrael is not rich in bodies of water that support fish (coll. *dāgâh*, Exod 7:18) and other water fauna, but there are adequate sources, including the Sea of Galilee, Lake Huleh, the eastern Mediterranean Sea, and many brooks and streams, including the Jordan River. The richness of edible water fauna is not expressed in the biblical sources, but the existence of a fish gate in Jerusalem (Zeph 1:10; Neh 3:3) strongly suggests that fish and other sea creatures were sold and consumed in inland regions. The Hebrew Scriptures do not mention any fish by name, and the dietary instructions state:

> *Of creatures that live in water these may be eaten: all,*
> *whether in salt water or fresh, that have fins and scales;*
> *but all, whether in salt or fresh water, that have neither*
> *fins nor scales, including both small creatures in shoals*
> *and larger creatures, you are to regard as prohibited. . . .*
> *Every creature in the water that has neither fins nor scales*
> *is prohibited to you.* (Lev 11:9–10, 12; see also Deut 14:9–10).[1]

While the dietary instructions are stated in very general terms, zooarchaeology and the study of water fauna in preindustrial Palestine can provide data enabling the reconstruction of water life in earlier periods. Biblical references and comparative studies of ancient neighboring cultures also help in reconstructing both fishing practices and the processing of the catch.

Fishing

Most fishing took place in the two large inland bodies of water, Lakes Kinneret and Huleh, or along the eastern shores of the Mediterranean Sea. It is safe to say that fishing also took place along the streams and rivers. However, while fish ponds are represented in Assyrian reliefs (Pritchard 1969b:fig. 114) and in Egyptian paintings (Bodenheimer 1960:70),[2] none are known from archaeological remains or any other sources in Iron Age Eretz Yisrael.

In antiquity, several methods were employed for catching fish; these are known both from biblical references and from ancient artistic representations in neighboring cultures.[3] A reference to fishing in Egypt appears in Isaiah 19:8, where the prophet speaks about "all who cast their hooks (*ḥakkâh*) into the Nile and those who spread nets (*mikmoret*) on the water. . . . "[4] Another biblical term for a fishing net is *ḥerem* (Hab 1:15–16; Ezek 32:3). According to Nun, both *mikmoret* and *ḥerem* are used for the dragnet—known also as seine—a type of net that is spread by boats (*sîrôt dûgâh*, Amos 4:2)[5] and needs two teams of fishermen (*dayyâgîm*, Isa 19:8) to drag each of its ends to the shore to bring in the catch (Fig. 6.1) (see also Bodenheimer 1935:432; Nun 1993).[6] The use of a seine is demonstrated by a model from Thebes that shows the net being dragged by two boats (Pritchard 1969b:fig. 109) (Fig. 6.2). Another illustration of a seine being pulled by two teams is seen on a painted relief from the tomb of Mereru-ka dated to the Sixth Dynasty (Pritchard 1969b:fig. 112). When the fishing is over, the nets need to be spread at a particular location (*mištaḥ ḥărâmîm*, Ezek 26:5, 14; 47:10) for repair and drying (Nun 1993:54).

A second type of net, the cast net (*rešet*), is used by one individual (Bodenheimer 1935:432; Nun 1993:53) and is referred to in Ezekiel 32:3.[7] A third type, the trammel net, is made of three layers, with the outer two having larger mesh than the middle one, thus causing fish that enter to get entangled between the layers. Nun suggests that the biblical terms *mĕṣôdâh* and *māṣôd* refer to this type of net, and that Ecclesiastics 9:12 describes its effects, "like fish caught in the destroying net (*mĕṣôdâh rā'âh*) . . . " (see also Job 19:6) (Bodenheimer 1935:432; Nun 1993:55).[8]

Fish spearing (Fig. 6.3), which was practiced in antiquity, is probably alluded to in Job 40:31 using the term *ṣilṣal dāgîm*.[9]

Though fishing was practiced throughout the Iron Age, there are very few direct references to Israelites engaged in fishing. Living off the riches of

Figure 6.1. *Fishing with a seine. A detail from Tomb 15 at Beni Hasan.*
(From Newberry 1894:pl.iv.)

Figure 6.2. *Fishermen using a seine. From tomb of Ra-hotep.*
(© BPK, Berlin, and the Staatliche Museen, Berlin.)

Figure 6.3. *Spear fishing. A detail from Tomb 17 at Beni Hasan.*
(From Newberry 1894:pl. XI.)

the sea is alluded to in Deuteronomy 33:19, when Zebulun and Issachar are told that "they will draw from the abundance of the sea. . . . " In the Song of Deborah (Judg 5:17) the tribes of Dan and Asher are connected to seafaring, and possibly fishing, "and Dan, why did he tarry by the ships? Asher remained by the seashore, by the creeks he stayed." Where did the Israelites learn fishing? Possibly from their neighbors the Phoenicians, who even during Nehemiah's time were responsible for selling fish in Jerusalem (Neh 13:16).

Processing the Catch

In Lake Kinneret, ten out of eighteen species are commercially important, the two most common of these are the barbel group (carps, Family Cyprinidae) (Fig. 6.4h–k), and the musht (*Tilapia galilea*, Family Cichlidae)(Fig. 6.4g).[10] The most common way of consuming freshly caught musht is by frying; the barbel, by boiling. Fresh fish must be consumed close, in time and space, to the place of the catch, because unprocessed fish spoil easily. Since fish bones and scales have been found at inland archaeological sites, it is safe to assume that the fish consumed there were first processed and then transported overland to market. There were several ways to process fish, including smoking,

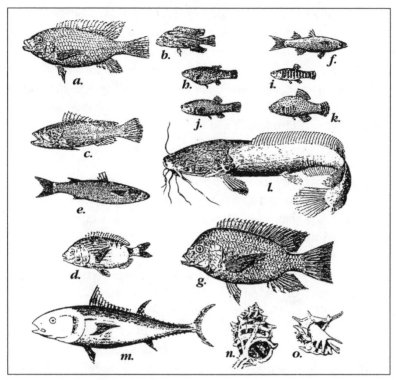

Figure 6.4. *Water fauna.* (From Bodenheimer 1935:pls. LVII, LIX, LXVI, LXVII, LXIX.)
Freshwater fish
  a. Mouth-breeder (Family Cichlidae), Tilapia cilli.
  b. T. flavii-josephi.
  f. Mullet (Family Mugilidae), Mugil capito.
  g. Musht (St. Peter's fish), Tilapia galilea.
  h. Barbel (carps; Family Cyprinodontidae), Cyprinodon
      sophia (female)
  i. C. sophia (male).
  j. C. cypris (female).
  k. C. cypris (male).
  l. Catfish (Family Clariidae), *Claris lazera.*
Marine fish
  c. Sea perch (Family Percidae), *Epinephelus aeneus.*
  d. Sea bream (Familiy Sparidae), *Sargus rondeletti.*
  e. Meager (Family Sciaenidae), *Sciaena aquila.*
  m. Mackerel (Family Scombridae), *Thynnus thynnus.*
Molluscs
  n. *Murex trunculus.*
  o. *Murex bandaris.*

drying, and salting.[11] The processing of fish (Fig. 6.5), probably for salting, is depicted in an Egyptian painting that shows two seated men slicing fish with long knives and flattening them (Brothwell and Brothwell 1969:fig. 20).[12] That the Egyptians were familiar with salting meat is illustrated in another early Egyptian painting showing three men, one cutting and cleaning a bird, another salting a bird, and the third putting birds (probably already salted) into tall storage jars (Brothwell and Brothwell 1969:162 and fig. 38).

Fish Remains and Their Significance
Analysis of fish remains can not only contribute to understanding the diet of a community, but also permits the study of fishing methods, the area and season of their exploitation, and a host of environmental conditions (Wheeler 1978:69).[13] To illustrate this, I would like to use the results of a study of fish bones found at two Late Roman-Byzantine castella—En Boqeq, situated on the west shore of the Dead Sea, and Tamar, further to the southwest. The study reveals that fish consumed at these sites originated in the Red Sea, the Mediterranean Sea, and freshwater bodies, most likely the Jordan River. Since the sites are distant from any of these points of origin, the fish must have been first dried or smoked and then transported (Lernau 1986).[14] Although this study is of a later period, the conclusions about the origin of the remains and their introduction into the site must be similar to those of what took place in the Iron Age.

The archaeological evidence for Iron Age fish consumption is more limited than what is available for meat-producing animals and is comparable with the scarcity of evidence for bird consumption. One of the sites where fish remains were recovered and studied is Jerusalem. Two excavated areas there, the Ophel and the City of David, yielded a large number of fish bones. At the Ophel, three families of marine fish and four of freshwater fish were identified (Lernau and Lernau 1989).[15] The marine fish originated in the Mediterranean Sea, while the freshwater fish, with the exception of one, the Nile perch (*Lates niloticus*)[16] can still be found in the rivers and lakes of Israel (Lernau and Lernau 1989:158). The Nile perch can reach large dimensions, and an Egyptian painting from Medum shows two men carrying a large Nile perch on their shoulders. The fish, its mouth tied on a pole, dangles between the men, the tail dragging on the ground (Brothwell and Brothwell 1969:fig. 21). Based on finds at other sites, it has been suggested that this fish might

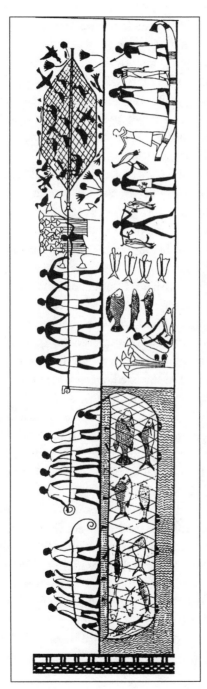

Figure 6.5. *Panel from Tomb 2 at Beni Hasan depicting fishing with seine (left), fowling in the marsh with a large trap (upper right), and fish processing and water fowl exchange (bottom right).* (From Newberry 1893:pl. XII.)

have been prevalent in waters near Jerusalem during the period in question, but its present absence from freshwater in Israel makes this suggestion hypothetical. One of the freshwater fish found in these sites, the Nile catfish, is considered non-kosher because of its lack of scales; thus it is very interesting how popular it was among Iron Age II Jerusalemites. Since the catfish's head is not edible, it was usually left where the fish was caught and processed, which may explain why it is the only specimen for which no head bones were discovered (Lernau and Lernau 1989:155).

The excavations at the City of David also yielded a large number of fish bones (215), the majority of which (183) came from Iron Age II loci. Most of the fish consumed in this period in Jerusalem, a little less than two thirds, were marine fish (Lernau and Lernau 1992:table 5).[17] The fish consumed at the City of David belong, generally, to the same families of fish consumed at the Ophel. These include gilt-head sea bream, white grouper, meager, Nile perch, Nile catfish, and flat-headed gray mullet.[18]

The variety of fish remains discovered in Jerusalem demonstrates that the inhabitants of the city had trade relations with several parts of the country. The freshwater fish most likely came from the Yarkon River or any of the other rivers on the coastal plain, or from the Jordan and its several tributaries. The marine fish originated in the Mediterranean Sea. No fish remains with origins in the Red Sea were found in these excavations. Again, it is important to note that since the Nile perch does not presently occur in Israel, it might have been imported to Jerusalem from Egypt. If the Nile perch was not native to Palestine and indeed was imported from Egypt, it might explain why the prophets Isaiah (Isa 19:8) and Ezekiel (Ezek 32:3) were so familiar with Egyptian fishing practices and metaphors.

The commercial relations between Eretz Yisrael and Egypt are underlined by additional data from other sites. Other inland sites also yielded fish remains exhibiting the presence of fish in the diet and indicating trade relations with other locations where the fish originated. Unpublished data[19] show that, during Iron Age I and II at Rosh Zait, in the Western Galilee, marine fish[20] and freshwater fish[21] were prevalent, together with remains of Nile perch (*Lates niloticus*). One bone of the latter fish was found at Iron Age I Tell Wawiyat, also in the Galilee. The discovery of Nile perch remains at such northern sites suggests long-range trade relations that stretched all the way to Egypt.[22]

Sites located on or near the Mediterranean coast tell a similar story. At Acco, on the northern coastal plain, fish remains from Iron Age I and II strata include remains of marine fish,[23] and freshwater fish,[24] as well as a large number of Nile perch bones. The large site of Ashkelon, on the southeastern Mediterranean coast, presents a similar picture. Here, too, strata dated to Iron Age I and II yielded remains of marine[25] and freshwater[26]fish, and a large number of Nile perch bones (O. Lernau, personal communication). Additionally, one fish bone of the Family Bagridae, originating in the Nile River, and a large number of bones of the Family Sparidae, originating in Lake Bardawil in northern Sinai, suggest strong trade relations with Egypt and the regions in between. Tel Gerisa and Tel Qasile, both located on the Yarkon River, also have remains of Nile perch.[27] All of these finds strongly suggest that during the Iron Age there was lively trade in fish with Egypt and northern Sinai, not only by landlocked sites but also by sites located near their own marine or freshwater fish resources.

Preliminary studies of bones from landlocked Tell Halif, in southwestern Judah indicate the presence of fish bones in the Iron Age II level dated to the time of the destruction of the city at the end of the eighth century B.C.E. (Arter 1995). All fish reaching this destination had to originate at distant sites located on the coast or near the Jordan River. Being closer to the coast than to the Jordan River strongly suggests that this was the area where the Tell Halif fish originated. This indicates that, even during the turbulent years at the end of the eighth century B.C.E., commerce was maintained between the hill country and the coast. This fact might have even deeper implications concerning the political climate and historical events of the period. Since Hezekiah, king of Judah, led an anti-Assyrian coalition that included some of the coastal Philistine cities, the fish remains found at the Judahite site of Tell Halif might have been delivered there by the coastal members of the coalition.

∼

While the Hebrew Scriptures have meager information concerning fish and their consumption, there are several references to fishing using precise terminology. Because of this, together with the abundance of fish bones at sites close to and far from fishing sources, it is safe to assume not only that the Israelites consumed a variety of fish but that some of them were also familiar with fishing. The variety of fish represented in the archaeological record is indicative of a lively trade in this commodity reaching as far as Egypt. However, the presence of certain varieties raises the question of adherence to the laws of prohibition. Another question that cannot be answered here is why no fish are mentioned by name, although their consumption is evident.

MOLLUSCS

Assorted Shells

Throughout the history of mankind, shells were collected for food, for decorative or utilitarian purposes, or for producing artifacts (Brothwell and Brothwell 1969:59). In certain communities located at the source of the shells, the meat of the molluscs was consumed before the shell was used or shipped away. Particular shells were used for decoration as beads, pendants and as other objects. For example, local marine shells of *Cardium rusticum* Linnaeus and *Glycemeris lividus* Reeve found in a Hellenistic context at Ashdod were perforated artificially, with some traces of coloring, strongly suggesting their use as ornaments. This use goes back to earlier times, as indicated by the occurence of such shells in earlier strata (Parotiz 1982). Shells were used for inlay in other media, such as wood, and for forming special features like eyes on clay figurines from Neolithic Jericho (Pritchard 1969b:fig. 221).

Most land snails recovered in archaeological excavations are local and of recent origin, however, remains of land snails dated to earlier periods could belong to specimens consumed by the inhabitants of the site. During the Pre-Pottery Neolithic and the Bronze Age in Jericho, terrestrial species such as *Helix prasinata* and *Levantina spiriplana* were used as food. Since no burning was detected, it appears that they were not roasted. Moreover, since the shells were intact, it indicates that the animals must have been killed before they were extracted, most likely by boiling. However, what vessel was used for this purpose in the Pre-Pottery Neolithic period is still a question (Brothwell and Brothwell 1969:64). That snail consumption was continued in the Iron Age is suggested by a group of seventeen terrestrial shells (genera *Sphincterochila* and *Helicella*) found in one locus in Beersheba (Hellwing 1984:110).

Shells recovered in archaeological excavations help in tracing trade connections between the find site and the source of the shells. One such site is Beersheba, where remains of the marine *Glycymeris violacenscens* indicate that this landlocked site had trade relations during the Iron Age I with either the Red Sea or the Mediterranean Sea (Hellwing 1984:110). The City of David in Jerusalem is another such site that yielded a large collection of shells traceable to their source. Being a large urban center with international connections, Jerusalem was the final destination of many exotic items and objects of luxury. Molluscs recovered in the Iron Age strata of the City of David came from freshwater and marine sources. The closest source was the streams and springs surrounding Jerusalem, where the *Melanopsis praemorsa buccinoidea*, a freshwater gastropod whose remains are represented in Iron Age strata, still lives (Mienes 1992:122–3). However, another freshwater specimen, also known from several other sites in Israel, is *Aspatharia rubens caillaudi*, an edible mussel found in the Nile River (Mienes 1992:126–7). The long duration of the trade relations that brought this shell to Jerusalem is evident because remains were found in strata dating from as early as the tenth century B.C.E. through the destruction of the city in 586 B.C.E.

Marine molluscs in the City of David came from two sources, one relatively close and the other quite distant. Three-fourths of the marine shells can be traced to the Mediterranean Sea, approximately 52 km away, and the rest to the Red Sea, approximately 270 km away (Mienes 1992:128–9). Although a few individual murex shells were found at the City of David, the very small numbers do not support the possibility of a dyeing activity at this location. Marine shells, including muricids, could have been used in the kitchen, as ornaments, and as building materials (Spanier and Karmon 1987:179–80).

Murex

Tĕkēlet (violet) and *ʾărgāmān* (purple) were two varieties of purple dye produced from the secretion of molluscs caught on the shores of the Mediterranean Sea.[28] The "purple snail" of the ancients was in reality a group of related molluscs. Those available on the eastern Mediterranean coast were *Murex brandaris* (Bodenheimer 1935:pl. LXIX:13), *Murex trunculus* (Bodenheimer 1935:pl. LXIX:11), and *Purpura haemostoma* (Spanier and Karmon 1987:pl. D).[29] Herzog suggests that *Janthina pallida* and *J. prolongata* (Spanier and

Karmon 1987:pl. H), both prevalent in the Mediterranean, were used for the production of *tĕkēlet*. (in Spanier 1987:139–42).[30]

Each individual mollusc (Fig. 6.4n–o) yields a very small amount of liquid from its hypobranchial gland, thus large quantities of shellfish are required for the commercial production of dye. In his *Natural History* (IX, 125–142), Pliny describes in detail the process of catching the whelk, extracting the raw material, producing the dye, and finally using it. Vitruvius (VII, 13. 1–3) complements this description. According to him, the purple that was most prized was obtained "from sea shells which yield the purple-dye. . . . When the shells have been collected, they are broken up with iron tools. Owing to these beatings a purple ooze like liquid teardrop is collected by bruising in a mortar" (Forbes 1956:116–7). Archaeological finds indicate that the Phoenicians, as well as the Greeks, smashed the murex shells to extract the secretion.[31] It has been suggested that when only small quantities of shells are found in archaeological context, they might have been used for food rather than for dyeing (Barber 1991:228).

The antiquity of the dye-making process is evidenced by a tablet found in 1933 in Ras Shamra (Ugarit) and dated to ca. 1500 B.C.E. that mentions the local dyeing of wool (Forbes 1956:112). Stieglitz suggests that dyeing with *Murex* molluscs was originated by the Minoans on Crete as early as ca. 1750 B.C.E. (Stieglitz 1994:53)[32] Nevertheless, the prophet Ezekiel considers Tyre a center of purple garments (*nopek ʾărgāmān*, Ezek 27:16) and of violet cloths (*gĕlômey tĕkēlet*, Ezek 27:24). Being precious, the dye was used in making royal clothing, such as "the purple robes worn by the Midianite kings. . . . " (Judg 8:26). There were other uses for dyed thread and textiles. As mentioned in Numbers 15:38, for identification and as a reminder, the Israelites were ordered, "Make tassels on the corners of your garments. . . . Into this tassel you are to work a violet thread (*pĕtîl tĕkēlet*)." The importance of *tĕkēlet* in ritual weaving caused later Jewish traditions to invest time and scholarship in either maintaining or recreating the process for future generations (Herzog in Spanier 1987:142).[33]

Murid dyes were used only with wool because they were not suited for linen (flax). The warm climate of the Near East fostered the common use of linen, which was dyed with other natural substances, rather than the use of wool, which could be dyed with the murid dye (Bridgeman 1987:159). Like other inhabitants of the region, the Israelites most likely preferred wearing light linen clothes made of locally grown or imported flax. Thus, dyed wool became a decorative element woven into lighter textiles or garments, and this was probably the reason for the violet thread woven into the fringes of Israelite clothes. However, a whole garment made of a mixture of wool and flax (*šaʿaṭnēz*, Lev 19:19) was prohibited from being worn. It seems that the woolen thread could only be worked into the tassels, as implied in Deuteronomy 22:11–12: "You are not to wear clothes woven with two kinds of yarns, wool and flax together. Make twisted tassels on the four corners of the garment which you wrap round you."[34]

Biblical traditions maintained that the Tabernacle was decorated with linens dyed violet, purple, and scarlet (Exod 26:1, 31, 36) and that priestly garments were also made of yarns dyed with these colors (Exod 28:5–8, 15 and more). One of the weavers in charge of the work for the Tabernacle was Aholiab son of Ahisamach, of the tribe of Dan (Exod 35:35; 38:23). Is it possible that, after its move to the north, Dan's proximity to the Phoenicians put the members of the tribe in contact with those artisans and the Danites learned some of the trade secrets of the Phoenicians?[35] Nevertheless, there are no biblical references to Israelite secular use of *tĕkēlet* and *ʾărgāmān,* and it seems that only late writings refer to the throne as covered by purple material (Song 3:10).

Evidence of the ancient dyeing industry is preserved in shell mounds and installations scattered throughout the Mediterranean basin (Stieglitz 1994).[36] Other data pertaining to this industry include traces of dye in clay vessels used in the process (Karmon and Spanier 1988:185). One coastal site with evidence of shells and clay vessels containing traces of dye is Tel Shiqmona, located south of present-day Haifa, where material dated to as early as the ninth century B.C.E. was uncovered (Karmon and Spanier 1988).[37] Another coastal site, Tel Dor, yielded an intact purple dye manufacturing installation dated to the Persian period. "The installation consists of a deep pit filled to the brim with crushed murex shells. A small channel leads to some sort of collecting vat. . . . The bottom of the channel was filled with a thick deposit of lime, with some residue of pigment still adhering to it. It seems that quicklime was used to extract the dye from the molluscs" (Stern

and Sharon 1987:208). These finds can be attributed to the fact that the area
was under Phoenician influence during these periods.[38]

Extra-biblical records illustrate the importance of dyed textiles, since
they were worthy of mention in tribute and booty lists. Shalmaneser III (858–
824 B.C.E.) records on his Black Obelisk that, in his eleventh year, he received
as tribute from Karparundi of Hattina "silver, gold, tin, wine, large cattle,
sheep, garments, linen" (Pritchard 1969a:280). Tiglath-Pileser III (744–727
B.C.E.) upon his return from his campaign to Syria and Palestine, reports of
booty that included "garments of their native (industries) (being made of)
dark purple wool . . . " (Pritchard 1969a:282). In the annals from his ninth
year, after a campaign to Syria-Palestine and Arabia, the booty included animals of different species and "linen garments with multicolored trimmings,
blue-dyed [takiltu] wool, purple-dyed [argamannu] wool. . . . also lambs whose
stretched hides were dyed purple" (Pritchard 1969a:283).

The Egyptians must have not been as intrigued by dyed garments as
the Assyrians; at least their records do not show such fascination. One of the
Egyptian reports after Thutmoses III's battle at Megiddo mentions that the
Pharaoh received "much clothing of that enemy" (Pritchard 1969a:238), but
the nature of the clothing is not clear. It is very possible that, having flax and
linen, the Egyptians did not see any need for woolen garments.[39]

~

Throughout history, terrestrial and marine molluscs were a dietary element
present in the archaeological record. After consuming the meat, some of the
shells were used for the production of objects such as jewelry and inlays.
Certain marine molluscs, Family Muricidae, were used in the production of
purple dye, mostly for wool and woolen garments. The small amount of raw
material available from each individual mollusk influenced the high price of
this dye and made it a luxury item. The purple dying industry was the spe-
cialty of the Phoenicians, thus the geographical close proximity of some of
the Israelites suggests the possibility that to a certain degree they were also
engaged in this activity.

Notes

1. Milgrom, on the basis of studies conducted by E. C. Haderlie of the Naval Post-Graduate School in Monterey, California, suggests that the Israelites were unfamiliar with fish because the eastern Mediterranean was a zoogeographical cul-de-sac, where there was no food to feed any fish. Accordingly, the nutritional silt coming from the Nile was flowing counterclockwise along the coast but in currents "too deep for most fauna to reach it until it surfaced in the Aegean Sea." This situation was changed with the opening of the Suez Canal in 1869 (Milgrom 1991:660). This does not explain several things, such as how the dunes were formed along the Mediterranean coast if the currents were too deep for providing food for the fish, the variety of fishbones found in excavations, or why the Israelites neglected to identify these fish by name. Whether the Mediterranean Sea was a "dead sea" (something I do not believe) or not, the Israelites ate fish and needed to identify them for themselves or for the fish traders.

2. The Egyptians also had ornamental fish ponds as part of their gardens. Ponds full of consumable fish were known from around the city of Per-Rameses. Some consumable fish (red *wedj*-fish and *bedin*-fish) could be found in the canals (Pritchard 1969a:471).

3. Catching fish by spear (or harpoon), hook and line, and nets is often shown in paintings in Egypt (Bodenheimer 1960:69; Carrington 1972:76) and Mesopotamia.

4. A good illustration of line fishing is shown in Tomb 3 at Beni Hasan (Newberry 1893, 1894:pl. xxix). Fishing with hook and line can be seen on a relief from Sennacherib's palace at Kuyunjik (Pritchard 1969b:fig. 114). In Mesopotamia, fishing was done by professionals using rods, nets (set or thrown), spears, harpoons, and baskets (von Soden 1994:89).

5. Another term for a boat or raft is *ṣinnâh* (*ṣinnôt*, Amos 4:2).

6. This fishing method can be seen in several tombs at Beni Hasan, especially Tombs 2, 3, 15, and 29 (Newberry 1893, 1894). The use of a seine is suggested in Matthew 13:47–48. A linen net from the second century c.e. was discovered in a cave near Ein Gedi (Nun 1993:55).

7. Ezekiel 12:13 and 17:20 use *rešet* in parallelism with *měṣûdâh*. For the cast net, see also Mark 1:16–8 and Matthew 4:19–20. Cast nets were found in Egyptian tombs of the second millennium B.C.E. (Nun 1993:53). For illustration of its use see Pritchard 1969b:fig. 112.

8. Nun suggests that the use of a trammel net is described in Luke 5:1–11 (Nun 1993:55). Fishing by traps is illustrated in Pritchard 1969b:fig. 112.

9. Using spears for fishing is illustrated in Tombs 3 and 17 at Beni Hasan (Newberry 1893, 1894:Pls. xi and xxxiv).

10. On freshwater fish in Palestine, see Bodenheimer 1935:422–33.

11. Dried and salted fish were an important food carried by the ancient

Phoenicians and Carthagenians on their long sea voyages (Bodenheimer 1960:70). The Assyrians reported receiving tribute from the Philistine cities that included processed fish, in Akkadian *dugla-ma-qar-te nuni* and *lat-tu nuni* (Elat 1977:253, table 14).

12. Processing fish is depicted in Tombs 2 and 29 at Beni Hasan (Newberry 1893, 1894:Pls. XII and XXVIII).

13. Examples of different kinds of fish are depicted on the walls of the temple of Queen Hatshepsut at Deir-el-Bahari. The illustrations also include other marine life (Carrington 1972:76). Marine life, including different kinds of fish, is illustrated on the north wall in chamber AI in the mastaba of Mereru-ka (ca. 2350–2190 B.C.E.) (see also Carrington 1972:16). Different types of fish and other marine fauna, such as crabs and eels, are depicted in Assyrian reliefs (Frankfort 1954:pls. 99–100,102a, 103).

14. One identifiable fish bone at Iron Age I Beersheba (Hellwing 1984:110) suggests that this landlocked site also had contacts with, at least, one fish producing community. Other Iron Age I landlocked sites where fish bones were found are Shiloh and 'Izbet Sartah (Hellwing 1984:112, table 18; Hellwing and Adjeman 1986). Iron Age II Lachish yielded one truncal vertebra of a large fish, possibly five years old, of the Serranidae (Grouper) family (Lernau 1975:90).

15. Freshwater fish: Nile catfish = *Claris gariepinus* (Fig. 6.4l); Nile perch = *Lates niloticus*; flat-headed gray mullet = *Mugil cephalus* (Fig. 6.4f) and thin-lipped gray mullet = *Liza ramada*; Mouth-breeder = *Tilapias* (this family includes St. Peter's Fish) (Fig. 6.4 a,b,g). Marine fish: white grouper = *Epinephelus aeneus* (Fig. 6.4c); gilt-head sea bream = *Sparus aurata* and common sea bream = *Sparus pagrus* (Fig. 6.4d); meager = *Argyrosomus regius* (Lernau and Lernau 1989) (Fig. 6.4e).

16. Egyptian trade in fish is recorded in the Wen-Amon story where the fish are delivered in baskets (Pritchard 1969a:28). The *Lates niloticus* was popular in Egypt, and since the Egyptians embalmed it, it received the same treatment as other popular animals (Bodenheimer 1960:128).

17. However, most of the bones (five out of six) from the Iron Age I belong to freshwater fish, Nile perch (Lernau and Lernau 1992:table 7).

18. During earlier and later periods, Jerusalemites consumed fish similar to those found here and at the Ophel (Lernau and Lernau 1992).

19. Personal communication by Omri Lernau.

20. These include *Epinephelus aeneus* (white grouper; Family Serranidae) and other fish of the Family Sparidae originating in Lake Bardawil in Northern Sinai.

21. *Clarias gariepinus* (catfish; Family Clariidae), which is common in all local streams and brooks.

22. The discovery of Nile perch remains (in addition to marine fish bones) at the landlocked site of Kuntilet 'Ajrud, on the border of the Sinai and the Negev (O.

Lernau, personal communication), should not be surprising because of the close proximity of the site to Egypt and its being on a major caravan route.

23. Among the marine fish were found remains of those belonging to the Family Sparidae, Family Scombridae, and *Argyrosomus regius* (meager; Family Scienidae).

24. These included *Clarias gariepinus* (catfish; Family Clariidae).

25. These included large number of bones of *Epinephelus aeneus* (white grouper; Family Serranidae), *Argyrosomus regius* (meager; Family Sciaenidae), and fish of the Family Sparidae, Family Scaridae, and Family Scombridae.

26. These included remains of the Family Cyprinidae (carp), and large numbers of bones belonging to catfish and Family Cichlidae (mouth-breeders). Fish bones of the Family Mugilidae cannot indicate their source, since these fish multiply in saltwater and then migrate and mature in freshwater brooks.

27. In addition, Tel Qasile yielded marine fish bones (O. Lernau, personal communication). A variety richer than all of the sites mentioned above was identified at Tell Hawam, at the mouth of the Kishon River. This site had only a small number of Nile perch bones (O. Lernau, personal communication).

28. Herzog has shown that both terms refer to dyes of molluscan origin (Herzog in Spanier 1987:140).

29. Forbes claims that when Pliny speaks about the "murex or baccinum," he means the *Purpura haemostoma* which, according to Forbes, occurs on the Atlantic coast only and was not available to the Phoenicians (1956:115). However, excavations in the eastern Mediterranean basin prove it otherwise (Karmon and Spanier 1988:184). For a good description of the muricid snails of the dye industry, see Spanier and Karmon 1987.

30. Mienis argues against the suggestion that the Family Janthinidae was used for the manufacture of *tĕkēlet*. He suggests that the muricid *Trunculariopsis trunculus* was the sources of this dye (Mienes 1987:204–205).

31. For a good and detailed description of the whole process, see Forbes 1956:112–21, Karmon and Spanier 1987:147–9, and Barber 1991:228–9.

32. This has been suggested also by others (Barber 1991:228).

33. Because they were very expensive, measures similar to the making of the biblical tassel were taken during the Roman and Byzantine periods for decorating garments, especially linen (Bridgeman 1987).

34. Joseph's tunic, sometimes referred to as "the coat of many colors," should be mentioned here. Several scholars translate this garment as "a long tunic" that reaches the ankles and the hand (Davies 1883:515; Brown et al. 1906:821; Buhl 1910:644; Koehler and Baumgartner 1953:768). A second suggestion is that it was a "tunic composed of variegated pieces" (Koehler and Baumgartner 1953:768), possibly each piece of different material or different color.

35. This makes perfect sense if the idea is accepted that the description of the Tabernacle is modeled after the Solomonic Temple and that workers in the Temple were supposed to be of similar backgrounds and traditions. This is further enhanced by the report in 2 Chronicles 2:12–13.

36. Unfortunately, broken shells are not collected in most excavations, thus it is impossible to verify the extant of a dyeing industry at these sites.

37. Karmon and Spanier review the archaeological evidence from several sites and relate it to information gathered from ancient sources (Karmon and Spanier 1987).

38. The stone vats discovered by Albright at Tell Beit Mirsim and identified by the excavator as dye vats have been proven to be olive presses, in spite of the fact that some scholars (Barber 1991:241–2) still consider them to be to the contrary.

39. It is obvious that the Egyptians did not ignore wool and woolen textiles completely, since ample evidence indicates their existence at least during late antiquity (Wilson 1933).

Chapter 7

Wild Animals

Wild animals were prevalent in Eretz Yisrael throughout the ages. Zooarchaeological investigations indicate that wildlife was rich and varied. Biblical references to wild fauna are numerous, as are those to hunting of several of these animals. Since early times, animals such as lions, wolves, bears, and foxes were hunted for their pelts (Clutton-Brock 1981:156). Other animals such as deer, gazelle, and birds were also hunted for their meat.

References to hunting, verified by zooarchaeological finds, reveal that this activity was not limited to the early days of the Settlement but continued even in the days of the prophets (Isa 8:14; 24:18; Jer 48:44; Hosea 5:1). Based on the number of mentions most hunting seems to have been done with traps and nets, though bow and arrows were also used (Gen 21:20; 27:3; Isa 7:24). Several biblical personalities are connected with hunting, but the prototype of all hunters was Nimrod: "He was outstanding as a mighty hunter—as the saying goes, 'like Nimrod, outstanding as a mighty hunter'" (Gen 10:9). Another famous biblical hunter was Esau (Gen 21:20; 25:28), and the story of how he lost his birthright and blessing to his brother, Jacob, while on a hunting mission for his father, Isaac, is well known (Gen 27). That Israelites were engaged in hunting for meat is attested by Levitical law: "Any Israelite or alien settled in Israel who hunts beasts or birds that may lawfully be eaten must drain out the blood and cover it with earth" (Lev 17:13).[1]

In the neighboring countries of Egypt and Assyria, hunting of particular animals was a royal sport in which the kings participated, and there are a number of monuments describing their prowess in words and art. One pharaoh, Thutmoses III, was quite a hunter. He hunted lions, wild cattle, elephants, and rhinoceros (Carrington 1972:74–4). Whether such hunting was also practiced by the kings of Israel and Judah is not known for lack of evidence, but it is quite possible since many wild animals were prevalent.

UNGULATES

Among the animals permitted to be eaten, Deuteronomy 14:5 mentions several that from other references seem to be wild and from their description belong to the ungulates (hoofed animals).[2] The list includes ʿayyāl, ṣĕbî, yaḥmûr, ʿaqqô, dîšôn, tĕʾô, and zemer, which are sometimes translated as "buck, gazelle, roebuck, ibex, white-humped deer, long-horned antelope, and rock-goat." Out of the last five, only two—yaḥmûr and tĕʾô—are mentioned once more in the Bible, in 1 Kings 5:3 and Isaiah 51:20, respectively. The other three are mentioned only in the Deuteronomic list. With this lack of information, it is very hard to offer a definite identification of species, but it can be determined that the Israelites were engaged in hunting wild animals for meat. The other two animals on the Deuteronomic list, ʿayyāl (fem. ʿayyālâh) and ṣĕbî (fem. ṣĕbiyyâh),[3] are mentioned several other times, three of which are elsewhere in Deuteronomy, where they appear together as animals that can be eaten. Together with the yaḥmûr, both also appear in the list of animals served on Solomon's table (1 Kgs 5:3). Both animals were known as fast running (Gen 49:21; Song 8:14), and they were considered beautiful to the degree that they became symbols of personal beauty (Prov 5:19; Song 4:5) or beauty in general (Isa 4:2; 13:19; Ezek 20:6). At times, the ṣĕbî is considered the symbol of the land (Ezek 20:15) and the people (2 Sam 1:19) of Israel.

Deer and Gazelle

The ʾayyāl may be identified with the roe deer (Cervus capreolus) or fallow deer[4] (C. mesopotamicus), since until recently both were very common in Eretz Yisrael. While the roe deer, which was common on Mount Carmel and in Upper Galilee, became extinct at the end of the last century or the beginning of the present century, the fallow deer, which was observed in the last century at Mount Tabor and Upper Galilee, could still be observed in the present century east of the Jordan (Bodenheimer 1935:114).

Bones of *Dama dama mesopotamica* were found at several sites, including in the Early and Late Bronze Age strata at Tel Kinrot (Hellwing 1988–89), in the early Iron Age strata at Beersheba (Hellwing 1984), at Iron Age II Lachish (Lernau 1975), and in the Persian period stratum at Tell el-Hesi (Bennett and Schwartz 1989). These finds show that the deer's use as a supplementary meat source was not limited to a certain period or region.

The ṣĕbî can be identified with several of the gazelle species (*Gazella gazella, G. dorcas, G. subgutturosa*), all of which are extremely common in hunting scenes on reliefs, seals, and paintings in Assyria, Egypt, and North Syria.[5] Each of the three occupies a particular niche, though they sometimes overlap; and their skeletal remains are not easy to tell apart in an archaeological context. The dorcas gazelle (*G. dorcas*) (Fig. 7.1) is the smallest and prefers the semidesert areas with acacia scrub.[6] The Arabian gazelle (*G. gazella*) is next in size and inhabits the mountains and foothills of Arabia, where the rainfall is higher. The largest of the three is the rhim or goitered gazelle (*G.*

Figure 7.1. *Gazelle,* Gazella dorcas. (From Bodenheimer 1935:pl. VII.)

subgutturosa) which besides its heavier build has differently-shaped horns in the male; the female lacks horns (Clutton-Brock 1981:169 and figs. 17.8–10). Some scholars suggest that the Egyptians kept gazelles as pets (Carrington 1972:79), and some maintain that these animals were probably domesticated to a certain extent in some regions during the Upper Paleolithic, Natufian, and early Pre-Pottery Neolithic, or even later (Salonen 1976:255; Legge in Clutton-Brock 1981:170). However, as Clutton-Brock demonstrates, these animals lack the traits that made other ungulates domesticable (Clutton-Brock 1981:170–1).[7] A wall painting in the tomb of Khnum-hotep III at Beni Hasan (ca. 1890 B.C.E.) depicts a group of Asiatics coming to Egypt leading two horned animals, one of which (the second in line) appears to be a gazelle (Pritchard 1969b:fig. 3) (see Fig. 3.3). The accompanying inscription does not shed light on the role or source of the animal, but it indeed looks tamed.[8] A similar type of gazelle, with horns bent somewhat forward (*Gazella subgutturosa*), is seen in a hunting relief from the palace at Nineveh in which a herd of males, females, and young, all depicted in a very realistic manner, are walking from the right side of the register to the left. One of the males looks back as if to see whether any danger is present (Frankfort 1954:fig. 113; Uerpmann 1987:fig. 46). Another hunting scene from Khorsabad shows a man aiming his bow into the air as if to shoot at birds, while another one walks away holding his bow and arrows and a third man, walking in the same direction, carries a dead gazelle over his shoulder (and a hare in his right hand) (Frankfort 1954:pl. 98). The gazelle looks very similar to the type mentioned above.

A gold plate from Ras Shamra (ca. 1450–1365 B.C.E.) (see Fig. 2.8), depicts people hunting gazelles (and bulls) with bow and arrow from a chariot with the help of dogs. (Pritchard 1969b:fig. 183). As depicted in artistic representations and mentioned in written records, hunting from a speeding chariot was quite common. Another scene, this time from Egypt, shows King Tut-ankh-Amen in his chariot chasing gazelles (and ostriches) and shooting them with his bow and arrows, while hunting dogs bite the hind legs and throats of some of the gazelles (Pritchard 1969b:fig. 190). Hunting for food with bow and arrows is specifically referred to in the Hebrew Scriptures, as in the well-known account about Esau, when his father Isaac told him, "Take your hunting gear, your quiver and bow, and go into the country and get me some game. Then make me a savoury dish, the kind I like, and bring it for me to eat so that I may give you my blessing before I die" (Gen 27:3–4).

It has been suggested that herds of gazelles were hunted with the use of "desert kites." These are installations made of two long walls, built of local stones, with a broad opening leading to an apex where a trap was laid. The animals were driven unsuspectingly toward the wide opening, and they were then caught in the trap at the other end of the installation. The earliest kites in the Negev were probably built during the Chalcolithic period, and similar methods were still utilized by the Bedouin during the last two centuries (Meshel 1974).[9] Gazelle bones were found at many sites, including the Iron Age I settlement of 'Izbet Sartah (Hellwing and Adjeman 1986), the early Iron Age strata of Beersheba (Hellwing 1984), the Iron Age II Lachish (Lernau 1975), the Persian period stratum at Tell el-Hesi (Bennett and Schwartz 1989), and several strata (including the Iron Age) at Tel Michal (Hellwing and Feig 1988) and Tell Halif (Zeder in Seger et al. 1990). These finds indicate that, during the biblical period, the gazelle served as a supplementary meat source at least in the central and southern parts of the country.

Ibex

Another native and common ungulate is the ibex (*Capra ibex*), which is actually a wild caprine (Fig. 7.2). There are several subspecies in Europe and Asia, all living in the high altitudes of the Italian Alps, Spain and Portugal, the Aegean Islands and Crete, the Himalayas and many other mountainous regions. The *C. Ibex nubiana* lives in the mountains of Arabia, southern Palestine, the Sinai, and southern Egypt. The ibex has untwisted curved horns with anterior ridges along the length of the horn and a goat beard (Bodenheimer 1935:pl. VII:2; Sanderson 1955:272; Bodenheimer 1960:50; Clutton-Brock 1981:59).[10] The ibex can be seen on many ancient monuments, seals, and seal impressions (Frankfort 1954:figs. 8A, 40C; Aynard 1972:fig. 20), metal and ivory objects (Frankfort 1954; Pritchard 1969b:figs. 290, 839), and more. Zooarchaeological remains of *C. ibex nubiana* are quite rare, mostly because it is difficult to distinguish between this and other members of the goat family. The most indicative osteological reference is that of the male horn-cores, such as those recovered in Early Bronze Age II Arad and Iron Age Hesbon (Uerpmann 1987:120). To a limited degree, ibex skin was used as parchment for writing documents, however it was not as common as goat skin (Oppenheim 1996).

Figure 7.2. *Ibex,* Capra nubiana. (From Bodenheimer 1935:pl. VII.)

Bovids

The subfamily Bovinae (wild cattle) of the bovid family[11] was represented in the ancient Near East by several species. The main members are the aurochs (*Bos primigenius*) or wild ox (*rĕʾēm*, Deut 33:17) and the wisent (*Bison bison*). Remains of water buffalo (*Bubalus bubalus*) have also been identified, but need confirmation (Uerpmann 1987:71). Being the ancestor of domestic cattle, the aurochs was well known in Egypt (Carrington 1972:74) where it was hunted by the kings. An early aurochs was found in the tomb of Sahure (Fifth Dynasty) (Bodenheimer 1960:123). One of the pharaohs who is known for such exploits is Amenophis III who hunted wild cattle (Carrington 1972:75). During the pre-dynastic period the bulls had curved horns with tips turned inward; they later acquired lyre-shaped horns with tips turned outward. During the reign of Thutmoses III, the aurochs was rare. When he heard of the presence of a herd, he immediately left Memphis and hurried by chariot through the night to the place indicated. There, in four days, he killed seventy-five out of a herd of 176 aurochs. This killing is commemorated on a special hunting scarab. The last document related to the aurochs is an impressive hunting scene of Rameses III chasing bulls by chariot (Bodenheimer 1960:123). Artistic representations show that aurochs were also present in Anatolia (Frankfort 1954:pl. 124), and zooarchaeological remains in Iron Age strata at several sites in Iran attest to its presence there. Only one Iron Age site in Jordan, Hesbon, has yielded the possible remains of this animal. Its presence in Eretz Yisrael is evident by its mention in Numbers 23:22; 24:8; Deuteronomy 33:17; and Psalms 92:11.

Hunting might have lowered the populations of both the aurochs and the wisent, but the latter has survived to the present. It seems that the main reason for the disappearance of the aurochs is the absorption of its young and females into domestic herds, causing a loss of genetic identity (Uerpmann 1987:72, 76).

The presence of wisent (bison) in the ancient Near East is known from artistic representations (Aynard 1972:fig. 23; Uerpmann 1987:fig. 30). However, close similarities of osteological characteristics of *Bos primigenius* and *Bison bison* strongly suggest that many specimens identified as aurochs may belong to the wisent. The ancient wisent of the Near East was a close relative of the now-extinct *B. b. caucasicus*. The shrinking of the wisent population in the Near East must be due to hunting as well as to strong competition with the aurochs and its domestic descendants (Uerpmann 1987:78). The wisent is a large, furry animal that slopes from the shoulders to the rump and has ox-like horns and a slight beard (Uerpmann 1987:fig. 29). A forest animal, it prefers coniferous forests, where it browses and grazes (Sanderson 1955:258).

The water buffalo (*tĕʾô*) is another bovid that has survived to the present (see Fig. 2.10d). The *Bubalus arnee,* progenitor of the domestic *Bubalus bubalus,* was known in the Near East from artistic representations, but only recently were zooarchaeological remains reported from a Halafian site in northern Syria. The habitat of the water buffalo is riverine forests and freshwater swamps, which have been shrinking and forcing the animal's retraction into very a limited area (Uerpmann 1987:78). In the first half of the present century, herds of domestic water buffaloes still inhabited the Huleh swamps and the marshes in the Sharon and coastal plain (Bodenheimer 1935:122).

The bovids (aurochs, buffalo, bison, and zebu) frequently appear on Mesopotamian monuments. While the aurochs is represented from earliest times onward, representations of buffalo are mostly limited to the Agade dynasty (second half of the third millennium B.C.E.) and are associated with Gilgamesh and his friend Enkidu (Aynard 1972:43, 45 and fig. 22).[12] The presence of water buffaloes in Eretz Yisrael is attested by the references to the *tô'* in Deuteronomy 14:5 and Isaiah 51:20.

OTHER HERBIVORES

Hyraxes and Hares

Other herbivores that lived in the ancient Near East include the *šāpān* (= hyrax, *Procavia capensis* or *P. syraica*) and *'arnebet* (hare, Family Leporidae)[13] both considered unclean for consumption (Lev 11:5; Deut 14:7).[14] The reasons for the prohibition are stated as "the *šāpān*, because though it chews the cud it does not have cloven hoofs . . . ;[15] the *'arnebet*, because though it chews the cud it does not have a parted foot . . . " (Lev 11:5). Although they do not belong to the same family, these animals were possibly placed together because they have a similar look and their voracious eating habits make them appear as if they ruminate.

The hyrax (Fig. 7.3) is a small mammal that has doubly cloven hoofs, but its feet are padded; it is somewhat related to the elephant. There are two kinds of hyraxes, one of which is prevalent in the Sinai, Palestine, and Syria.[16] This group lives in large groups of up to eighty individuals (see the description in Ps 104:18) and protects itself by living in the crevices of rocky terrain (see description in Prov 30:26) (see also Sanderson 1955:291-92; Stevenson and Hesse 1990:27). Hyraxes are diurnal and herbivorous, but do not chew their cud. The reports of recent attempts to domesticate this animal illustrate that it was always a potential meat source (Stevenson and Hesse 1990).

The hare (or rabbit) is another small furry, herbivorous animal with a large appetite; "the name hare is preferred for the larger, long-legged, large-eared types that leap (the genus *Lepus*), while rabbit is reserved for the smaller, short-legged species that run (*Sylvilagus*)" (Sanderson 1955:113). Clutton-Brock distinguishes between hares and rabbits on the basis of their behavior (hares being more solitary) and in living above ground (rabbits burrow)

Figure 7.3. *Hyrax,* Procavia syriaca. (From Bodenheimer 1935:pl. vii.)

(1981:146). Their method of digestion might be responsible in part for their inclusion among the prohibited animals. After consuming large quantities of fresh green food, the animal excretes the partially digested food near its home and, after some time, eats it again. This process might be repeated more than once (Sanderson 1955:114; Clutton-Brock 1981:145). Hares have been hunted and raised for their meat and pelts.

Elephants and Other Large Animals

The elephant is not mentioned in the Hebrew Scriptures, but it needs to be included here for several reasons. Two large animals are referred to by the name elephant, the African Loxodon (*Loxodonta africana*) which is the largest living land mammal, and the Indian Elephant (*Elephas maximus*) which is smaller. The African elephant has larger ears and both sexes have very long tasks, while only the Indian male has tusks. Both animals evolved during the Pleistocene while other earlier species became extinct (on this animal see Sanderson 1955:292–4). Literary and artistic evidence suggest very strongly that elephants were prevalent in the Near East, including Syria, in early historic periods. Thutmoses iii hunted 120 elephants on his campaign in Syria, a fact recorded by his biographer Amen-en-heb and confirmed on the Barkal Stele (Carrington 1972:75). It is assumed that they belonged to the Indian species (Clutton-Brock 1981:114).

The middle panel of the Black Obelisk of Shalmaneser iii depicts animals brought as tribute from the country Musri. In addition to Bactrian camels and monkeys, the tribute includes "a river ox (hippopotamus), a *sakea*-animal (rhinoceros), a *susu*-antelope, elephants . . . "[17] (Pritchard 1969b:290–1 and figs. 351–5). Man's first attraction to the elephant was for its meat, but ivory was the reason later; only in the last four thousand years have elephants been tamed as beasts of burden.

Elephant ivory was used throughout the Ancient Near East, and carved ivories were found in Knossos, in the Minoan palace at Zacro and in Mycenae (Clutton-Brock 1981:117). In Syria-Palestine, ivory was carved and used for jewelry, boxes, inlay pieces, handles, figurines, and many more items (see for example Crowfoot and Crowfoot 1938). At several sites, ivory carvings were found starting in layers dated to the Chalcolithic period, especially in the southern region of Israel. While some of the carvings were made of hippopotamus tusks, at this point it is next to impossible to differentiate between the carved ivory of the two elephant species and that of the hippopotamus.

In the ninth-eighth century B.C.E., ivory carvings were luxury and prestige items, and examples of such were found in the royal palaces in Nimrud (Mesopotamia), Salamis (Cyprus), Arslan Tash (Syria) (Fig. 7.4), and Samaria (Israel).[18] They were all carved following the Phoenician ivory-carving style (Mazar 1990:503–504). The royal palaces with their ivory decorations (*hēykalēy*

Figure 7.4. *Cow suckling a calf. Ivory inlay, Arslan Tash.*
(Musée du Louvre/Antiquités Orientales.)

šēn, bātēy haššēn) are mentioned in the Bible several times (Amos 3:15; Ps 45:9). According to the Bible, in his palace Solomon "also made a great throne inlaid with ivory and overlaid with fine gold" (1 Kgs 10:7; also 2 Chr 9:17) and Ahab was singled out for "the palace he decorated with ivory . . . " (1 Kgs 22:39) where the nobility lolled "on beds inlaid with ivory . . . " (Amos 6:4).

Elephants have never been domesticated, and one possible reason is their long period of gestation, up to twenty-two months, which does not afford control over their breeding. However, throughout history individuals were captured, tamed, and trained to perform certain work or entertainment tasks, or used as "war machines." The latter use is recorded for the later historical periods. Darius, king of Persia, used tamed Asian elephants in his battle against Alexander the Great, and the Carthaginians under Hannibal used tamed African elephants during the Punic Wars (Sanderson 1955:294; Clutton-Brock 1981). In the Hellenistic period, elephants were also tamed for military use in Egypt (Carrington 1972:75), and the Seleucids used tamed elephants in their war against the Maccabees (2 Macc. 13:15) and others (1 Macc. 8:6).

The hippopotamus (*Hippopotamus amphibius*) is a very temperature-sensitive animal that was probably mentioned in the Hebrew Bible (as *běhēmôt;* Job 40:15) and on the Black Obelisk mentioned above. Hippopotamus remains were found at Bronze and Iron Age sites in the southern Levant, suggesting its existence there. However, it is possible that these finds are "trophies" derived from animals hunted in Egypt (Uerpmann 1987:46), where it inhabited the Nile all the way to the Mediterranean (Sanderson 1955:246). These animals caused much damage to crops, and their hunting in Egypt for ivory is recorded until the closing years of the pre-Christian era (Carrington 1972:72–73). Hippopotamus tusks were used for carving during the Chalcolithic period and Early Bronze Age, as finds from the Beersheba region indicate.

The origin of the rhinoceros mentioned in the Black Obelisk is hard to determine. No remains of this species younger than 30,000 years have been found in the Near East (Uerpmann 1987:13). At the present, there are four genera, two in Africa and two in Asia, one of which has only one horn made of congealed hair and not of bony material (Sanderson 1955:241–42).

CARNIVORES

Many carnivorous mammals were prevalent in Eretz Yisrael during biblical times. Although they were not exploited like the domestic animals mentioned in this work, their presence was an integral part of the landscape and needs a brief treatment.[19]

Lions

The lion (*'ărî* or *'aryēh*)[20] and the leopard (*nāmēr*) were the largest carnivores inhabiting the land. From the number of references to this animal, it appears that the lion was more common than the leopard, however both were feared as ferocious animals (Prov 22:13). The lion was considered the most ferocious (Judg 14:18) because of its roar (Amos 3:8) and the fact that it attacked people (Amos 5:19).[21]

The lion (*Panthera leo* or *Felis leo* L.) that inhabited the ancient Near East is now extinct; it belongs to the Family Felidae which includes also the jaguar, cheetah, tiger, leopard, panther, and lesser cats. While biblical references suggest that the lion was a most ferocious animal, modern zoologists describe it as "terrified of small children, and flapping laundry on a line . . . ." Furthermore, "although hunters, they may even lie down back-to-back with antelopes during the day, and they appear to make little more than one kill per month on an average" (Sanderson 1955:161). Most of the hunting is done by the female, and lions generally do not harm people unless startled, wounded, bothered, or driven by disease, extreme hunger, or some strange dereliction that affects whole populations of mostly unmated juveniles. The areas preferred by lions are savanna, grassland, and semiarid regions very similar to the conditions prevalent in parts of ancient Palestine (Sanderson 1955:161).

Biblical references suggest that the lion was prevalent in all parts of the country, including Bashan (Deut 33:22), Moab (Isa 15:9), the Jordan Valley (Jer 49:19; 50:44), and the territory of Judah (1 Sam 17:34–37). The lion was prevalent in the mountains (Song 4:8), in the forests (Jer 12:8; Amos 3:4; Micah 5:7), in the thicket (Jer 4:7), near Bethel in Judah (1 Kgs 13), and in the Shephelah near Timna (Judg 14).

One famous biblical encounter between man and beast is that of Samson who killed a young lion (*kĕpîr ʾarāyôt*) when he was on his way to visit a Philistine woman in Timna (Judg 14). According to biblical accounts, another person famous for killing a lion was Benaiah, son of Jehoiada, one of David's thirty heroes, "who once went down into a pit and killed a lion on a snowy day" (2 Sam 23:20). The Bible recounts that another lion slayer was David, but this is only according to his own testimony (1 Sam 17:34–37).

The herders considered the lion quite a menace because it attacked their herds. There are several references to such incidents (1 Sam 17:34–37; Amos 3:12) and it was depicted in art, as in the Samaria ivories, which show a lion attacking a bull (Crowfoot and Crowfoot 1938:pl. x:1–2). Therefore, on the Day of Judgment, when peace will reign, the prophet Isaiah dreams that "the calf and the young lion will feed together . . . and the lion will eat straw like cattle" (Isa 11:6–7).

Ancient art is replete with images of lions—they appear on orthostats and gate posts, as large and small statues; in reliefs, wall paintings, statuettes and figurines; and in cylinder seals and seal impression (for example see Van Buren 1930:pl. xli:196–9, xlii:200–5; Crowfoot and Crowfoot 1938:pl. ix; Steindorff 1946:pls. lix:311–4, lc:325b, 328, lxv:345a).

Hunting lions was considered a royal sport, and many hunting scenes were commissioned by the ancient kings. In Mesopotamia, lion hunting is depicted as early as the beginning of the third millennium b.c.e. in Uruk (Pritchard 1969b:fig. 182) and in Agade (Aynard 1972:fig. 22). Most famous are the reliefs from the Assyrian palaces, especially those of Ashurnasirpal at Kalah and Ashurbanipal at Nineveh (Hall 1928:pls xvii, xlvii–xlix; Aynard 1972:47–9). In both palaces, the hunters are shown using bow and arrow from their chariots, helped by men (possibly soldiers) armed with lances and other weapons. Many of the lions are seen lying dead with arrows piercing their bodies, while others are still trying to attack the chariot (Pritchard 1969b:fig. 184; Aynard 1972:fig. 24). Other famous reliefs from Nineveh are those of a lioness wounded by several arrows still roaring at her hunter (Frankfort 1954:pl. 111a) and of a dying lion hit with one arrow between his shoulders (Frankfort 1954:p. 111b). Lions were kept in parks for the king's pleasure (Frankfort 1954:pl. 108a),[22] and caged lions were used to provide the king with the thrill of the hunt when they were released toward him for target

Figure 7.5. *Ashurbanipal's lion hunt, a relief from his North Palace at Nineveh.*
(Aynard 1972:48.)

shooting (Hall 1928:pl. LI; Frankfort 1954:pl. 108: B). One relief from Nineveh
(Fig. 7.5) shows the king on foot killing a charging lion with his sword, piercing the animal through (Frankfort 1954:pl. 109).[23] Some reliefs show attempts
at taming lions (Hall 1928:pl. LII).

Lion pelts were worn in religious ceremonies; an Assyrian relief depicts a man wearing a lion hide with the head placed over his and the rest of
the skin laying down his back all the way to his ankles (Frankfort 1954:pl.
94A). The cultic importance of the lion throughout the Near East is underlined by its inclusion among the animals flanking the procession street at
Babylon (Frankfort 1954:fig. 33; Pritchard 1969b:fig. 762), by the discovery
of skeletal remains in cultic places, and by the animal's inclusion in cultic
scenes. These are discussed in chapter 8.

Egyptian pharaohs also considered lion hunting a royal sport, and they
enjoyed it as much as hunting wild ox (Carrington 1972:74). Egyptian pharaohs, like Assyrian kings, hunted lions from chariots with bow and arrow.
One such scene depicted on a casket from Tut-ankh-Amen's tomb shows him
in hot pursuit after lions, being helped by dogs and servants (Carrington
1972:fig. 36). Thutmoses III is credited with killing in the desert "seven lions

while out shooting in the twinkling of an eye" (Carrington 1972:74). Furthermore, it seems that Rameses II and Rameses III kept lions as pets in a semidomesticated state (Carrington 1972:80).[24] Were the kings of Israel and Judah also engaged in hunting lions? There is no recorded answer to this question, but this is very feasible since they had all the necessary ingredients and the ambition to be considered as great and powerful as the other kings of the ancient Near East.

The Hittites (Frankfort 1954:figs. 50, 87, 89 and pls. 128A, 133A), the Hurrians (Frankfort 1954:pl. 140), the Arameans (Frankfort 1954:pls. C–D),[25] and the Phoenicians (Frankfort 1954:pl. 169A) all employed images of lions in their art, whether in stone,[26] metal or ivory.

Glyptic evidence for the existence of lions in Palestine is available but limited. From Canaan, the most famous example is a basalt relief from Stratum IX of Beth-shan (fourteenth century B.C.E.) showing in the upper register a dog and a lion fighting, standing on their hind legs facing each other (Fig. 7.6). The lower register depicts a dog biting the rump of a lion (Pritchard 1969b:fig. 228). Canaanite Hazor also yielded a basalt orthostat shaped like a lion (Pritchard 1969b:fig. 856). A lion, together with other animals, is painted on the ewer from Canaanite Lachish (thirteenth century B.C.E.) (Pritchard 1969b:fig. 273). Lion images also appear on an ivory box and an ivory comb from Canaanite Megiddo (ca. 1350–1150 B.C.E.), and on several ivory carvings from Israelite Samaria (Pritchard 1969b:figs 67, 128–30). Israelite cultic objects are represented by a steatite incense spoon (ninth century B.C.E.) shaped like a lion's head, with the bowl of the spoon depicting the lion's lower jaw (Pritchard 1969b:fig. 592). Possibly the most famous lion image from ancient Israel is that of an eighth-century B.C.E. seal from Megiddo portraying a roaring lion with an inscription: "(Belonging) to Shema, servant of Jeroboam (II)" (Pritchard 1969b:fig. 276).

Lion skeletal remains from Palestine are very rare. Most notable is the skull found in Jaffa on a floor of a structure, possibly a temple, dated to the end of the thirteenth or beginning of twelfth centuries B.C.E.. Half a scarab seal with the name of Queen Tiy, wife of Amen-hotep III, was found near the skull's teeth. This building might have been a temple in which a lion cult was practiced (Kaplan and Ritter-Kaplan 1993). Other sites where lion bones were found are Dan and Tel Miqne-Ekron (Wapnish and Hesse 1991:47). At Dan, a thumb bone of a lion with cut marks was found in a room adjacent to the

Figure 7.6. *Lion stele depicting two scenes of a struggle between a dog and a lion, from Beth-shan.* (Pritchard 1969:fig. 228.)

Israelite High Place. Laish, another name for lion and Dan's former name, possibly reflects the inclusion of lions in the cult. Judging from place names, other sites where a cult involving lions could have taken place were ha-Kephirah and Beit Leba'ot (Kaplan and Ritter-Kaplan 1993:656, 658). Several lion bones were found in Field I at Tel Miqne-Ekron (Hesse, personal communication). The nature of the structure in which they were found is still to be determined.

Leopards

The leopard (*Panthera pardus* or *Felis pardus cf tulliana*) is a close relative of the lion, but biblical references mentioning it are very few, suggesting that it was not as common. However, unlike the lion this animal is still in existence in the Near East; in nineteenth and early twentieth centuries it was seen in the Carmel and around Jerusalem, the Jordan Valley, Araba, Mount Tabor, and Jordan (Bodenheimer 1935:114). In more recent years, it has only been sighted in the Judean desert, especially in the area surrounding Ein Gedi.

The leopard (*nāmēr*) hunts at night and will attack humans as well as animals, preferring domestic mammals and birds. In action, it is quite fearless, and in the Bible it was considered the fastest mammal (Hab 1:8). The color of the leopard tends to reflect its habitat, thus those from dry, open, rocky, and treeless areas are inclined to be large and pale-colored. Those that live in damp, forested regions tend to be smaller, with larger black spots on a darker basal color (Sanderson 1955:162). The leopard's preferred habitat in Eretz Yisrael was the rocky mountainous regions (Song 4:8), and its spots were its trademark (Jer 13:23). The leopard's ferocity was legend. The prophet Isaiah foreseeing the Day of the Lord as a time of peace and tranquillity when natural enemies will live at peace with each other, declared the impossible, namely that "the leopard (will) lie down with the kid . . . " (Isa 11:6).

Illustrations of leopards in antiquity are very few. One of the earliest depictions is known from Shrine E VII 44 at Çatal Hüyük, dated to the Neolithic period, where the leopard was worshipped as the goddess "mistress of animals" (Cole 1972:36–8). An orthostat at Tell Halaf is decorated with a leopard with a collar (Bodenheimer 1960:100), suggesting that it was tamed. The leopard's ferocity is shown on a cylinder seal from the Early Dynastic III (middle third millennium) depicting a leopard attacking a horned animal (Pritchard 1969b:fig. 678). Leopard skins were probably used for making clothes and for covering certain objects, such as the soundbox of a harp shown in a wall painting from a tomb in Thebes (Thutmoses IV, ca. 1421–1413 B.C.E.) (Pritchard 1969b:fig. 208).

Bears

The Syrian brown bear (*Ursus syriacus*; Heb. *dob*) was quite common in Palestine during biblical times (Sanderson 1955:pl. 122). In general, bears do not molest humans unless they are surprised or attacked, and then they are more dangerous than any of the great cats. They can climb trees, swim, and cross any kind of terrain. Bears eat almost any animal or vegetable, and particularly like fish and sweet things, especially honey. In the winter they hibernate for long periods. Before leaving their winter quarters, they give birth, often to twins or triplets, which are very small. The cubs follow their mother for many months (Sanderson 1955:201).

The Hebrew Bible contains several references to bears. It is described as a prowling animal (Prov 28:15) with a distinct growl (Isa 59:11), lying in wait for its prey (Lam. 3:10), and chasing it until it is caught (Amos 5:19). Bears were considered dangerous, especially those who lost their young (*dob šakkûl*, 2 Sam 17:8; Hos 13:8; Prov 17:12). The Bible tells us that David, as a shepherd, had to fight away a bear (as well as a lion) to protect his herd (1 Sam 17:34–37). One incident involving bears is vividly described in 2 Kings 2. After receiving the prophetic leadership from Elijah, Elisha stayed in Jericho where he performed some miracles. "From there he went up to Bethel, as he was on his way, some small boys came out of the town and jeered at him, saying 'Get along with you, bald head, get along.' He turned round, looked at them, and cursed them in the name of the Lord; and two she-bears came out of a wood and mauled forty-two of them "(2 Kgs 2:23–24). Believed to be a ferocious animal, Isaiah mentions the bear grazing and living in harmony with the cow as a picture of peaceful coexistence in the End of Days (Isa 11:7).

Ancient Near Eastern art seldom depicted bears, possibly because they were not as visible as other animals. One Middle Assyrian cylinder seal (second half of the second millennium B.C.E.) shows a bear (with other animals), possibly at the entrance to a cave, pursued by hunters (Aynard 1972:50 and fig. 20). Another well-known depiction of a bear is part of a shell plaque from the soundbox of a lyre found at the king's grave in Ur (first half of the third millennium B.C.E.). The bear is standing facing an animal, possibly a donkey, playing a lyre (Aynard 1972:fig. 32).

Sightings of Syrian bears were reported near Tiberias, near Beth-shan, in the Golan and the Hermon during the mid-nineteenth and early twentieth centuries. They were dark brown and no taller than 140 cm. However, the bear population declined in the Hermon and Anti-Lebanon because the animal was hunted by German officers during World War One (Bodenheimer 1935:114). Bears were hunted in the region even in the Iron Age, as seen in an eighth-century B.C.E. relief from Karatepe that shows a hunter with a bow and arrows facing an animal resembling a bear (Pritchard 1969b:fig. 971). There is no evidence, written or otherwise, to show that bears were hunted in Eretz Yisrael during the Iron Age.

CANIDS

Wolves

The wolf (*Canis lupus*; Heb. *zě'ēb*) is the progenitor of the domestic dog, and is the indigenous dog of the sub-Arctic, temperate, and desert areas of the Northern Hemisphere (Sanderson 1955:196). Wolves are nomads, living either as individuals, in a family or in a pack. Except during the breeding season, they travel to follow their source of food and can adapt to different climatic conditions (Sanderson 1955:196).

The Hebrew Bible considers the wolf a fierce animal of prey: "Benjamin is ravening wolf: in the morning he devours the prey, in the evening he snatches a share of the spoil" (Gen 49:27).[27] Therefore, peaceful coexistence, when "the wolf will live with the lamb . . . " (Isa 11:6; see also 65:25) is a great metaphor for the End of Days. Ancient illustrations of wolves are very rare. One such depiction appears on a north Syrian Aramaic amulet dated to the seventh century B.C.E. (Bodenheimer 1960:44).[28] An earlier representation of a head of a wolf is known from the Jemdat Nasr period (Bodenheimer 1960:100). Osteologically, it is very hard to distinguish between skeletal remains of the wolf and those of dogs or jackals. Wolves are still prevalent in the Middle East, and at certain times they have been known to attack domestic animals as big as cattle and cause much damage.

Foxes

The fox (*Vulpes palaestinus*; Heb. *šû'āl*) and the jackal (*Canis aureus*; Heb. *tan*) are small canids, and thus are relatives of the domestic dog. Foxes are widespread, but they are not as smart as is traditionally believed. They are burrowing animals that kill all kinds of game, especially rodents. A big-eared fox (*V. familicus*), found all over the Middle East, lives on jerboas, gerbils, lizards, and insects (Sanderson 1955:197). The Song of Songs considers the fox a menace to vineyards, where it destroys the grapes (Song 2:15).[29]

Although the fox was never domesticated or tamed, the Bible relates an interesting story concerning Samson and three hundred foxes that he captured and tied in pairs; he attached a burning torch to the tails of each pair, sending them free to burn the Philistines' fields as revenge (Judg 15:4–5). Unfortunately, this legend has no basis in reality.

The fox is well documented in Mesopotamian art (Aynard 1972:43) as well as in Sumerian literature (Aynard 1972:65). Its association with the jackal made the fox a symbol of desolation: "Your prophets, Israel, have been like šû'ālîm among ruins" (Ezek 13:4). The ancient affinity of the fox with death is suggested by the placement of its skull, together with other objects, in the Neolithic shrine vii, 35 in Çatal Hüyük (Cole 1972:38).

Jackals

Jackals (Heb. sg. *tan;* pl. *tannîm*), which are divided into four distinct kinds and numerous races, subspecies, and other types, occupy the subtropical and tropical regions of Asia and the drier parts of Africa; they are still prevalent in large numbers in the Middle East. They are dog-like omnivorous scavengers (Ps 63:11), smaller than wolves, and look much like foxes. Jackals are nocturnal pack hunters that sleep in holes during the day. They interbreed with domestic dogs and wolves (Sanderson 1955:197). At night, while hunting, they make a distinct wailing sound (Micah 1:8) that announces their presence. Historically, they tended to live in ruins, thus the Bible associates them with destruction. Isaiah, as well as Malachi (Mal 1:3), uses jackals several times as a metaphor for desolation (Isa 13:22; 34:13; 35:7; 43:20). Predicting future destruction, Jeremiah says, "Listen, a rumour comes flying, a great uproar from the land of the north, an army to make Judah's cities desolate, a haunt of *tannîm*" (Jer 10:22). And the same fate is predicted for Hazor, "Hazor will become a haunt of *tannîm*, for ever desolate, where no one will live, no mortal make a home" (Jer 49:33).

In artistic representation, as in osteological remains, sometimes it is very hard to tell the difference between a jackal and a dog or fox. What appears to be a jackal carrying a table laden with animal heads and meat, is seen on the end of a lyre soundbox from Ur (Pritchard 1969b:fig. 192).[30] The jackal was given magical powers. A jackal, or a wolf, devouring a child appears on a soft gypsum amulet from Arslan Tash dated to the sixth or seventh century B.C.E. (Pritchard 1969b:fig. 662). This object was supposed to carry good luck in child birth. In the Egyptian pantheon, the god Anubis was portrayed in the form of jackal or as a human with a jackal head (Pritchard 1969b:figs. 548, 639). The jackal's popularity in Egypt is demonstrated by jackal (and dog) heads that were attached to the game pieces of a gameboard found in the tomb of Ren-seneb in Thebes (Pritchard 1969b:fig. 213). Jackals play also an important role in Egyptian literary genres such as fables (Carrington 1972:87).

OTHER ANIMALS

Hyenas

The striped hyena (*Hyaena hyaena*; Heb. *ṣābôʿă*), the more common representative of this family (Hyaeninae), is quite common in the Near East even today. The spotted hyena (*H. crocuta*) was known in antiquity but only as an exotic animal (Bodenheimer 1960:44). It is commemorated in the Bible in the locational name *gēy ṣĕbōʿîm*, "Valley of Hyenas" (1 Sam 13:18),[31] somewhere in the vicinity of Jerusalem, possibly Wadi Qelt. The striped hyena is identified by several black upright stripes on its sides and horizontal stripes on its legs. Its front legs are much longer than its hind legs and its tail is short and bushy; it feeds on carrion, and even digs up human burials. Although it lives in holes, whenever they are available, it prefers caves or ruins. Unlike the spotted hyena, it drags its food into the lair. All hyenas have very strong jaws that can crack large bones such as ox leg bones (Sanderson 1955:175–76). Hyenas were prevalent in Egypt during Early Dynastic period (Carrington 1972:70), and although they are not mentioned in the Bible as living animals, everything suggests that they were extant.

Exotic Animals

Written documents and artistic representations reveal that the Egyptian pharaohs and the Assyrian kings collected exotic plants and animals and kept them in parks established especially for this purpose. At least one the Israelite kings, Solomon, is recorded as having been engaged in a similar enterprise. "The king had a fleet of *taršîš*-ships at sea with Hiram's fleet; once every three years this fleet . . .came home, bringing gold and silver, ivory,[32] apes, and *tukkiyyîm*" (1 Kgs 10:22 = 2 Chr 9:21).[33] The term *tukkiyyîm* appears only in this context. It has been translated as "monkeys" (Suggs et al. 1992), "poultry or baboons" (Holladay 1978), or "peacocks" (Brown et al. 1906; Mandelkern 1967; Leteris 1984). Whatever the term stands for, it is obvious that this must have been some exotic specimen. Examples of exotic animals being brought to kings are depicted on the Black Obelisk (Pritchard 1969b:figs. 353–4), where apes[34] and an elephant are shown as being brought to Shalmaneser III, and in a wall painting in Deir el-Bahri showing apes (and different plants and animals) brought from Africa to Queen Hatshepsut (Davidson 1965:50).

Another biblical term that describes an exotic (or unfamiliar) animal is *taḥaš/těḥāšîm*, which appears quite a few times (Exod 25:5; 26:14; 35:7, 23; 36:19; 39:34; Num 4:6, 8, 10–2, 14, 25; Ezek 16:10). It has been translated as "badger" (Davidson 1848; Leteris 1984), "dugong" (Brown et al. 1906; Suggs et al. 1992), "seal" (Davies 1883), "porpoise" (Koehler and Baumgartner 1953), "porpoise or dolphin" (Holladay 1978), and some other animals like "giraffe" (Mandelkern 1967). Whatever this animal is, the term is always used in reference to its hide, which was used in the construction of the Tabernacle. Only once (Ezek 16:10) does it appear in a different context: "I gave you robes of brocade and sandals of *taḥaš*-hide; I fastened a linen girdle round you and dressed you in fine linen." The implication is that this type of hide is an expensive and luxurious item.

I dare suggest that the reference is to the crocodile (*Crocodiles vulgaris* or *C. niloticus*) which is native to Egypt. It was a sacred animal; the god Sobek is often shown as one, and the god Horus appears standing on crocodiles (Steindorff 1946:734–44). Nevertheless, it was hunted from small boats, as seen in hunting scenes, and was mummified in great numbers. Crocodiles were sent as gifts to foreign kings; one was sent during the Twenty-first Dynasty to Tiglath-Pileser I (Carrington 1972:73). Although now extinct there, crocodiles were prevalent in Palestine. Crocodile sightings near Caesarea were recorded in the Roman and Crusader periods, and during the nineteenth and early twentieth centuries they were seen at several places, around the Kishon and between the Carmel and Yarkon River (Bodenheimer 1935:197).

~

The Israelites were well familiar with wildlife, as is evidenced by the count-less biblical references that refer not only to the animals but also to their habits. Wild animals occupied a place in Israelite daily life, as did domestic animals. Several of the ungulates were hunted for their meat and other by-products. Other animals were hunted for their pelts or tusks, and some animals were hunted just for sport. Special expeditions were sent to distant lands to bring back exotic animals that were considered luxury items and were displayed or kept as pets. Some wild animals even found a place in the cult, either as an object of worship or for sacrifice.

Notes

1. That hunting was an ancient way for provisioning among the Israelites is suggested by the fact that the word for hunting, *ṣayid*, also means "food" in general (Job 38:41; Prov 12:27; Neh 13:15).

2. Their description in Deuteronomy 14:6 states that they are hoofed, have cloven hoofs, and chew their cud.

3. The young of both animals are known as *'oper* (Song 2:9, 17; 4:5; 7:4; 8:14).

4. There are several subspecies of fallow deer (*Dama Dama* or *D. mesopotamica*) that were prevalent in the Near East in the past and can be identified with the biblical *'ayyâl*. The deer is also represented in the Samaria ivories (Crowfoot and Crowfoot 1938:pl. x:8, 8a).

5. It should be noted that the dama gazelle (*G. dama*) which is rare on ancient Egyptian monuments (Bodenheimer 1960:49–50), now lives in the Sudan, as does *G. soemmeringi*, which also appears on old Egyptian monuments (Bodenheimer 1960:123). Recent investigations show that *Gazella gazella* was the dominant gazelle in the Levant until after the PPNB period, when *Gazella dorcas* appeared from northeast Africa and replaced it in most regions (Tchernov et al. 1986/87).

6. The Egyptians, who mummified different kinds of animals, also embalmed ungulates. One collection includes twenty gazelles (*Gazella dorcas*), which were more common in Lower Egypt, and *G. arabica*, more common in Upper Egypt) (Bodenheimer 1960:127). For *Gazella dorcas* see Bodenheimer 1935:pl. vii:3.

7. The same conclusion was reached in a recent study, which demonstrated that, in spite of certain claims, no domestication of gazelles was achieved in the Natufian period (Dayan and Simberloff 1995).

8. The possibility that the Egyptians owned tamed wild ungulates can be surmised from a fragment of a tomb relief, possibly from the New Kingdom, showing an attendant leading an antelope by its horns, which he holds with both hands (Steindorff 1946:fig. 270).

9. Meshel, following I. Aharoni, suggests that *pokeret haṣṣēbāyîm,* which appears as a family name in Ezra 2:57 and Nehemiah 7:59, is the biblical term for "kite" (Meshel 1974:135).

10. Another member of this family is the wild goat (*C. aegargus*) (Uerpmann 1987:113–8).

11. "The bovids . . . comprise all the ruminants which grow permanent horns from large wild cattle down to the tiny duiker antelopes of Central America" (Uerpmann 1987:68).

12. See also the Akkadian seal impression depicting two heroes, each with a water buffalo (in Frankfort 1954:45D). The list of tribute from Musri on Shalmaneser III's Black Obelisk lists "a river ox," sometimes translated 'hippopotamus' (Pritchard 1969b:281). Can this reference be to water buffalo?

13. In Akkadian, *annabu* (Brown et al. 1906:58).

14. The *šāpān* is also mentioned in Prov 30:26; Ps 104:18.

15. Although the *šāpān* was considered unclean, during monarchical times several individuals were named after this animal.

16. The tree hyrax is prevalent in other regions, such as the forests of Africa.

17. Clutton-Brock suggests that these were Syrian elephants (1981:fig. 11.9).

18. Ivory is referred to in the Hebrew Bible as *šēn* (1 Kgs 10:18) and *šenhāb* (1 Kgs 10:22; 2 Chr 9:21).

19. That carnivores were common in the land is alluded to in the Joseph story. The plot that Joseph's brothers concocted to convince Jacob of his favorite son's disappearance could only have been hatched in this context (Gen 37:31–33).

20. A second term for lion is *lābî* (Gen 49:9 and other; fem. *lĕbiyyā*, Ezek 19:2), a third term is *layiš* (Isa 30:6) and a fourth one is *šaḥal* (Job 4:10). What difference between these terms indicates is hard to tell. A lion cub is *kĕpîr* (Isa 31:4) or *gûr* (Deut 33:22).

21. Judah was likened to a young lion (Gen 49:9).

22. See also Register IV on the Black Obelisk, in which are shown two lions in a park with a stag between them being attacked by the lion behind (Pritchard 1969b:fig. 351, 355).

23. A good presentation of these reliefs can be found in Hall 1928:pls XVII, XLVII– LII.

24. Syrians bringing tribute, including a tamed lion, to Tut-ankh-Amon are seen in a painting in the tomb of Huy (Pritchard 1969b:fig. 52). The goddess Qadesh, and possibly other goddesses, was depicted standing on a lion flanked by other gods, worshippers, or cultic symbols (Pritchard, 1969b:figs 470–74, 830). Similar depictions of gods standing on a lion or other animals (e.g., bull) are known from other parts of the Near East (Pritchard 1969b:figs 486, 500–501, 522, 531, 534, 537).

25. A beautiful, two-thirds of a mina bronze weight in the shape of a lion from the Shalmaneser V (726–722 B.C.E.) palace in Nimrud is inscribed in Aramaic, "Two-thirds [mina] of the land" and in cuneiform: "Palace of Shalmaneser (V), king of Ashur, two-thirds mina of the king" (Pritchard 1969b:263–64 and fig. 119).

26. The body of a winged lion is a major element in sphinx images in many cultures (Pritchard 1969b:figs. 644, 646, 648–50, 765).

27. See also references in Jeremiah 5:6; Ezekiel 22:27.

28. Wolves can be seen also on Roman mosaic floors from Palestine (Bodenheimer 1960:44).

29. It is quite possible that the reference is not to foxes, but to the fruit bat (Borowski 1987:156–7). The term *sĕmādār,* which is sometimes translated as 'blossom' is most likely a reference to a kind of grapes (Borowski 1987:104).

30. Animals performing human tasks are depicted not only in Mesopotamian art, but also in that of other cultures. One such example from Egypt is a satirical drawing on papyrus of animals replacing people carrying out various roles (Carrington 1972:fig. 39).

31. See also Nehemiah 11:34 and possibly Genesis 14:2.

32. Although the term indicates "ivories" is it possible that live elephants were brought back?

33. The term *taršîš* has been translated as a place name or a type of ship.

34. Monkeys appear in Mesopotamia on amulets, terracottas, and bas-reliefs. Since they were not native to Mesopotamia, they must have been brought there from the Indus Valley or Egypt. One ivory from Nimrud depicts what may be a Nubian bringing a tribute of an antelope and a monkey (Aynard, 1972:57, and fig. 29). Monkeys kept as pets in Egypt were brought there from south of Egypt (Carrington, 1972:79).

Chapter 8

Animals in the Cult
of Ancient Israel

The sacrifice of animals was a long-standing custom in the Ancient Near East, and evidence for such practices can be found in biblical and extrabiblical sources and in the archaeological record.[1] The Israelites considered animal sacrifices to be Ancient rites, and this concept is supported in the traditions that place such sacrifices in the early days of humanity in general (Gen 4:4; 8:20) and of the nation in particular (Gen 15). Since much has been written about the role animals played in the cult of Israel and her neighbors, in this section I will summarize some of the major points related to the topic and paint a broader picture animals' role played in the cult in light of what has been presented in the previous chapters. This will include discussions of the different animals used in the cult, their remains found in cult centers, and some special occasions that were celebrated by particular sacrifices. In doing so, our understanding of the place of animals in the daily life of Ancient Israel will become clearer and more complete.

The close relationship animals had with the people of the Ancient Near East can be seen from their depiction in association with the gods. Gods are depicted as standing on top of animals (Fig. 8.1), as well as in the shape of animals; other images show them as creatures with either a human body and an animal head, or a human head an an animal body. Such presentations are not limited to non-Israelite cults, because YHWH, too, is likened to an animal several times, such as in the expression *'ăbîr ya'ăkob*, "the bull of Jacob" (Gen 49:24; Isa 49:26; Ps 132:2, 5) or *'ăbîr yiśrāēl* (Isa 1:24).

The bull (*pār*), the symbol of power and fertility, is mentioned in the Hebrew Scriptures numerous times, the majority of them in connection with

Figure 8.1. *Goddess on lion.*
(©BPK, Berlin, and the Ägyptisches Museum, Berlin.)

sacrifices (Num 7). However, at times the Bible uses the bull as a symbol for YHWH, employing the term *'abbir*. In several cultures in the Ancient Near East, the bull was the symbol of various gods (Pritchard 1969b:figs. 828 [from Ras Shamra], 832 [from Hazor]),[2] a fact which must have influenced Israelite iconography and found expression verbally and visually. The story of the Golden Calf (Exod 32) is one instance of this influence. Another example is that of the golden calves erected in Bethel and Dan (1 Kgs 12:28; 2 Kgs 10:29). The bronze bull statuette from the Bull Site in the Samaria hill country (Fig. 8.2) is a remnant of such cultic use of the bull in early Israelite religion. Architectural remains uncovered at this site included a large stone circle with a *maṣṣēbâh* (standing stone) on its eastern side, where the statuette was found. Mazar, the excavator, suggests that "the

Figure 8.2. *Bronze bull from the Bull Site.* (Mazar 1990:fig. 9.13.)

figurine was probably used by Israelite settlers in this region of the northern Samarian hills" (1990:352).

In examining the role animals played in the Israelite cult, the analysis should consider domestic and wild animals separately, and each of these categories needs an additional division between mammals and birds.

In line with general Near Eastern tradition, the largest group of animals associated with the Israelite cult is the domestic ruminants—goats, sheep, and large cattle.[3] Members of this group served as the actual sacrifices offered on different occasions. There are biblical prescriptions for when and how such animals should be sacrificed, but here I would like to examine the animals themselves.[4]

The animals to be sacrificed were large cattle (*bāqār*) and small cattle (*ṣoʾn*), adults as well as firstlings (Lev 1:2; Num 7), male as well as female (Lev 1:10; 3:6; 5:6). To be sacrificed, an animal cannot have "any defect or serious blemish, for it would be abominable to the Lord your God" (Deut 17:1; see also Lev 22:21).[5] A first-born ruminant must be sacrificed and, unlike other animals, cannot be redeemed (Exod 34:19; Num 18:17; Deut 12:6).[6] To be sacrificed, the ruminant must be at least a week old (Exod 22:29; Lev 22:27). Specially fattened animals (*kārîm*, Isa 34:6; *mēḥîm*, Ps 66:15; *mĕrîʾ*, 1 Kgs 1:9) were also sacrificed. However, since not everyone could afford to sacrifice a ruminant, two turtledoves or two pigeons could be substituted for it (Lev 5:7; 12:8). The source of these birds must have been domestic because otherwise it would have been very hard to secure their blemish-free suitability for sacrifice.

Certain parts of the animals were considered choice and could either be sacrificed or contributed to the priests for their consumption. The priests received the shoulder,[7] the cheeks, and the stomach (Deut 18:3; see also Exod 29:26–27). *Ḥēleb* (fat) could not be eaten by Israelites (Lev 7:23), but had to be offered to YHWH. The fat was taken from the sheep's tail (*ʾalyâh*) and from other parts (Exod 29:22; Lev 3:9; and others). Furthermore, the blood was considered the soul of the animal and was part of the sacrifice (Isa 1:11; 34:6; Ezek 39:18; Ps 50:13).

Certain rituals required particular animals. One such ritual was the Red Heifer ritual which is described only once in Numbers 19. This ritual was supposed to provide "the water of purification" for people as well as objects.[8]

The prescription calls for "a red cow without blemish or defect, one which has never borne a yoke" (Num 19:2). The latter part of the verse suggests very strongly that the cow is an ordinary one that otherwise would have been pressed into service on the farm or somewhere else. In chapter 2, I suggested that the best candidate for this ritual was the Beirut cow which is reddish in color and is used for work. Another ritual is that of the Broken-necked[9] Heifer, which was done to expiate for a murdered person and is described in Deuteronomy 21:1–9. In this case, the animal had to be "a heifer that has never been put to work or worn a yoke" (Deut 21:3). Again, the animal must have been a member of the domestic herd, young enough not to have been used in any type of work.

Another definite requirement of certain sacrifices was that the animal be a yearling, which is the age when the animal reaches its optimal weight without too much investment in feed having been made. Several times the prescription is very definite that the sex of the animal be a male (Lev 9:3) or female (Lev 14:10) yearling. Sometimes the prescription calls for male and female yearlings (Num 6:14). At times the requirement is for a *kebeś* (male sheep) or *śĕʿîr ʿizzîm* (male goat) yearling (Num 7), and at other times it is for a *śeh* (young ruminant) (Exod 12:5) yearling. At times the sacrifice specifies an *ʿēgel* (calf) yearling (Lev 9:3) while at other times it calls for a *pār* (bull; Lev 16:3). The call for the offering of yearlings necessitates some loss of income because, in a meat producing economy, the herder tends to sell his surplus lambs at that age. However, the requirement also favors the herder because he needs to cull the herds and this presents an opportunity to fullfil an obligation to the deity without having to sacrifice beyond a certain amount. The sacrifice of an older animal would have demanded additional investment because of its extended care and feeding.

On special occasions, male adult animals were offered. These included he-goats (*ʿattûd*) and rams (*ʿayil*), sometimes purposely fattened (*ʾēyl milluʾîm*), so portions could be given to the priests (Exod 29:26–27; Lev 8:29). Bulls (*pār*) and oxen (*śôr*) were also sacrificed on designated occasions. Fattened bulls (*mĕrîʾ*) were sacrificed during festive events such as coronations (1 Kgs 1:9).[10]

THE PASSOVER SACRIFICE

One cannot escape the conclusion that the Passover sacrifice is particularly related to herding and animal husbandry. Israelite religion "did not create all its practices out of nothing, but adapted to its purposes many existing forms and conventions which it imbued with a new spiritual meaning" including the prebiblical Passover (Haran 1985:224). The prescriptions recorded in the Hebrew Bible have a very long history, and the present detailed description of the Passover celebration (Deut 16:1–8) is possibly the result of Josiah's reforms (Haran 1985:145–6). Elements of the biblical Passover can be traced to an earlier Feast of the Barley Harvest (Haran 1985:296–7), which suggests that the feast evolved as a combined celebration that probably reflects the union between the farmer and the herder.[11] This union makes Passover extremely important; and, to sanctify the *maṣṣôt/pesaḥ* affair, its action is placed in Egypt.

Here I would like to look only at the pastoral elements of the feast, those related to herding and animal husbandry. Some of the components of the feast can only stem from a pastoral background and can only be understood as such (Rylaarsdam 1962:668; Licht 1971:517–18; Bokser 1992:760). Both Wellhausen and Haran are right in suggesting that the origin of Passover is pastoral; but unlike Wellhausen, Haran is correct in proposing that the origins of the feast and sacrifice are not in the firstling sacrifice (Haran 1985:324–6). I follow the suggestion that the feast originated to celebrate the beginning of the *raḥil*, which occurs at the commencement of spring and can be compared to present-day celebrations that take place at the beginning of a season such as the blessing of the hounds at the beginning of hunting season, the blessing of the ships at the beginning of the fishing season, and other such events.

The time of the celebration is in the middle of the lunar month Abîb and not at its opening, when one would expect it to take place. Had it been at the beginning of the month the celebrants would have faced a dark night, but a full moon is necessary for the proceedings. Since the instructions call for a sleepless night (*lēyl šimmūrîm*, Exod 12:42) during which certain activities take place, the feast can only be done under a full moon. This feast is unique in Israelite cult because the Passover sacrifice is the only ritual calling for the sacrifice to be completely eaten at night.

The sacrificial animal had to be selected in advance to insure its being without blemish (Exod 12:4).[12] It had to be a yearling male signifying the first major culling of the herd. The animal had to be slaughtered "between dusk and dark," as the moon was rising. The participants "must take some of the blood and smear it on the two doorposts and on the lintel of the houses in which they eat the victims" (Exod 12:7).[13] Smearing or sprinkling of the blood was done with "a bunch of marjoram" (Exod 12:22). Since this was the symbolic beginning of moving the herds from the permanent site to the temporary grazing area (transhumance), the blood sprinkling must have been done for good luck. The participants closed themselves in the house after sprinkling the blood of the sacrifice on the doorposts to ward off bad spirits before and during the *raḥil*. Going out of the house before sunrise might have brought bad luck.

Other elements of the celebration enhance its relationship to the *raḥil*. Similar to the reasons given in the biblical prescriptions for the historical feasts—sitting in booths to imitate living conditions in the desert, for example, or eating *maṣṣôt* to relive the night of the departure from Egypt—the original observation of this feast intended to emulate the *raḥil*. By reenacting the whole event beforehand, the observation of the feast relied on sympathetic magic to influence the outcome and bring good luck during the wandering season. Participants had to be fully clothed at night (Exod 12:11) ready to start the march at daybreak. To induce good luck, during the night there must have been reenactments of the main components of the *raḥil*.

The Passover sacrifice had to be roasted on fire, not eaten raw or boiled, and it had to be consumed with unleavened bread and bitter herbs (Exod 12:8–9), thus duplicating the diet eaten by the herders during their wanderings. Boiling, rather than roasting, the kid (Deut 16:7) reflects a late version of Passover celebration more in tune with the observant urban society, introduced no later than the time of Josiah, as stated in 2 Chronicles 35:13: "They cooked the Passover victims over the fire according to custom, and boiled the holy offerings in pots, cauldrons, and pans, and served them quickly to all the people."

Eating the sacrifice completely, without saving anything for the morning of the next day is another sign that the ritual was marking the beginning of the *raḥil*, since nothing of the old can be transported to the new location. The participants in the Passover sacrifice were not allowed to leave the house until morning, because daybreak was the mark of a new season.

~

The Passover celebration and sacrifice are a reflection of the pastoral roots of Israel. Although the feast of Passover started as a family sacrifice on a special occasion in the middle of the month of Abîb, under the Deuteronomist it became a highly institutionalized affair in which the sacrifice still maintained a central position. The occasion celebrated the opening of the transhumance season, and the proceedings seem to reenact the activities that would take place during the wandering (*raḥil*). Many of the customs appear to be connected with sympathetic magic trying to influence the successful outcome of the grazing season. When the cult was centralized in Jerusalem in Josiah's reign and the urban population became more influential in cultic matters, some of the herding-related customs, such as roasting the sacrifice, were replaced, and the victim began to be boiled. However, the pastoral-rooted festival continued to be celebrated, possibly taking on a new meaning.

WILD ANIMALS AS SACRIFICES

In addition to domestic ruminants, the biblical record permits eating wild ungulates (Deut 14:5), but does not include them in the list of sacrifices. This group includes *'ayyāl, ṣĕbî, yaḥmûr, 'aqqô, dîšôn, tĕ'ô,* and *zemer*. The *'ayyāl* may be identified with the roe deer (*Cervus capreolus*) or fallow deer (*C. mesopotamicus*), and the *ṣĕbî"*with several of the gazelle species (*Gazella gazella, G. dorcas, G. subgutturosa*). The identification of the other animals is doubtful (see chapter 7).[14] Remains of wild ruminants have been identified among the bones recovered from cult places, raising the question of whether they were sacrificed there. If these animals were part of the cult, the question of their availability needs to be adressed. How were they caught without being harmed so they could be sacrificed, or were certain specimens kept as tamed animals?

Fowl were another faunal element offered as sacrifices, though not as widely as the ruminants. When a fowl was sacrificed, the instructions specified that it had to be a turtledove or a dove (Lev 1:14) There were two conditions under which birds were offered, either to replace a ruminant or when they were specifically designated. Certain sacrifices permitted or required the offering of fowl. When a leper[15] was to be declared pure (healed), or an object or house infected with fungus was purified, the ceremony, which has two stages, is described in detail (Lev 14); but while the type of birds to be offered in stage one is not specified, the prescription calls for two birds (*šĕtēy ṣippărîm*; Lev 14:4, 49). In stage two, however, a number of ruminants has to be offered as part of the purifying procedure; but when the offerer cannot afford it, he is required to bring two turtledoves (*šĕtēy tôrîm*) or two doves (*šĕnēy bĕnēy yônâh*, Lev 14:22, 30). Based on the latter reference, it is safe to assume that the call for two birds in stage one is actually for doves or turtledoves.[16] Another case when two turtledoves or two doves can be substituted for a she-ruminant is in the *ʾāšām* sacrifice (Lev 5:7). However, in this case, if the person is too poor and cannot even afford fowl, a tenth of an *ʾēypah* of *solet* (high grade flour) can be substituted (Lev 5:11). This allusion suggests very strongly that doves and turtledoves were raised under controlled conditions, because if they were in the wild, most everyone could afford them.

There are other instances where fowl were prescribed as sacrifice. When Abraham was instructed to select certain animals for a covenantal sacrifice (Gen 15:9), he was told to bring a turtledove and a *gôzāl* (a young bird).[17] Purification offerings called for two turtledoves or doves, as in the cases of a man with "a discharge from his private parts" (Lev 15:2–16) and when "a woman has a prolonged discharge of blood not at the time of her menstruation . . . " (Lev 15:25). A turtledove or a dove (in addition to a yearling) is prescribed for sacrifice by a woman who gave birth (Lev 12:6). However, if she could not afford the yearling, a turtledove or a dove could be substituted for the lamb (Lev 12:8). The large number of doves and turtledoves required for a variety of sacrifices suggests very strongly that these birds were raised in a domestic environment or under other controlled conditions.[18]

ANIMAL SACRIFICES IN BIBLICAL TRADITIONS

All of the above demonstrates that the prescriptions for sacrifices are quite precise, thus it can be instructive to examine some of the events when sacrifices where offered to see how the recorded traditions treat the sacrifices and whether they follow the prescriptions. The earliest mention of animal sacrifice in the Bible is the offering by Abel "who brought the choicest of the firstborn of his flock" (Gen 4:4). Noah, following the flood, offered as thanksgiving "beasts and birds of every kind that were ritually clean . . . " (Gen 8:20). The main difference between these two early events is that Abel brought domestic caprids, while Noah is said to have brought ritually clean animals and birds, which could have been wild animals. Abraham, the eponymous father of the nation, was told to offer as ratification of the covenant with YHWH "a heifer three years old, a she-goat three years old, a ram three years old, a turtledove, and a young pigeon" (Gen 15:9). This is the only precise description of a Patriarchal sacrifice, and the only case where three-year-old animals are prescribed for a sacrifice. The offering here is possibly so costly because of the momentous event. That young, rather than adult, animals were usually sacrificed can be seen from the following example. When Isaac was going to be sacrificed by Abraham and his life was spared, a ram was sacrificed in his stead. However, on the way to the Land of Moriah, Isaac inquired about a young ruminant (*śeh*) needed for the sacrifice, thus suggesting very strongly that this was the standard of the day, which was later replaced by a more costly adult ruminant (*'ayil*) (Gen 22). According to biblical records, the Patriarchs did not sacrifice many animals. Nevertheless, one such occurrence is referred to in vague terms when Jacob moved to Beersheba (Gen 46:1), and similar references are made to Moses and Jethro (Exod 18:12; 24:5). Aaron's sacrifices during the affair of the Golden Calf are also described in imprecise terms (Exod 32:6). A much more precise description, one that includes bulls and rams, is provided for the sacrifices offered by Balaam the Seer from Moab during his attempt to curse the Israelites (Num 23).

The big difference is seen in the traditions related to events that took place in the land during and after the Settlement. While there is a certain vagueness concerning the sacrifices in the desert, and it continues even when an altar is erected by Joshua on Mount Ebal (Jos 8:30–31),[19] with Gideon (Judg 6:25–28) and Manoah (Judg 13:19) we get a precise description of the animals offered.[20] A much more descriptive approach, which includes not only an enumeration of the animals sacrificed (1 Sam 1:25) but also the procedure followed by the priests (1 Sam 2:13–16), starts with Samuel's biography. It continues with the return of the Ark from Philistia (1 Sam 6:14) and with Saul's visit to Samuel (1 Sam 9:12–24; 10:3). The anointment of the kings opens new opportunities for sacrifices of animals; these are first mentioned in general terms (1 Sam 11:15; 13:9; 15:15) then become more specific (1 Sam 16:2). The establishment of the monarchy gave rise to the custom of sacrificing great numbers of animals, and this is described not only in general terms (2 Sam 24:25), but also with some precision (1 Kgs 1:9, 25). The practice was refined by Solomon when the Temple was inaugurated (1 Kgs 8:5, 62–64; 2 Chr 5:6), and according to biblical records, it was maintained throughout the monarchy, as attested to by the prophets (Isa 1:11; Jer 6:20; Ezek 43–46; Amos 5:21–22). Events similar to the one performed by Solomon were supposedly duplicated by Hezekiah (2 Chr 29:20–24, 32–35; 30:24) and Josiah (2 Chr 35:7–9) as part of their cultic reforms.

EVIDENCE OF ANIMAL SACRIFICES IN CULT CENTERS

Almost none of the Israelite cult places that have been located allow any verification of the biblical record. Nevertheless, the osteological remains recovered in the few excavated cult centers demonstrate adherence to biblical prescriptions and reveal that the majority of the animals sacrificed in the Iron Age were ruminants.

An Israelite/Judahite temple dated to the Iron Age II was discovered at the fortress of Arad, on the eastern edge of the Beersheba valley. The temple was in use from the tenth through the seventh century B.C.E. (Herzog et al. 1987).[21] Not much has been written about the osteological material found in the temple and around the sacrificial altar. However, a mention is made that in the pre-temple phase, dated to the eleventh century B.C.E., the site had a high place where "near the altar there were pits with burnt bones and the burnt skeleton of a young lamb" (Aharoni 1967b:270). The latest phase of the temple (beginning of seventh century B.C.E.) yielded two small stone altars flanking the entrance into the holy of holies. On top of each, inside shallow depressions, were found traces of animal fat (Aharoni 1967a:247, f.n. 30). Much burnt bone residue was uncovered, but details of the discovery have not yet been published.

An Iron Age I cult center identified by the excavator, A. Mazar, as Israelite and known as the Bull Site was discovered in the Samaria hills (Mazar 1982). Unfortunately, no bones were recovered from the site. However, another site in the general region and from the same period discovered on Mount Ebal furnished better results (Zertal 1986–87). This site (Figs. 8.3 and 8.4), which has been identified by its excavator, A. Zertal, as the altar built by Joshua (Deut 27:2–8; Jos 8:30–35),[22] yielded a large number of bones, some of which were burnt. Out of a total of 2,862 bones, 770 (27%) could be identified. Of the identifiable bones, 741 (96%) belong to the following: sheep and goats (65%), cattle (21%), and fallow deer (10%). In addition to these, 29 bones (4%) were identified as belonging to hare (*Lepus capensis*), marbled polecat (*Vormella pregusna*), an unidentified small carnivore, hedgehog (*Erinaceus europaeus*), carapace fragments of a tortoise (*Testudo graeca*), starred lizard (*Agama stellio*), an unidentified reptile (possibly a snake), and mole rat (*Spalax ehrenbergi*). Also identified were bones of partridge (*Alectoris* sp., possibly chukar partridge), and rock dove (*Columba livia*), as well as a bone of a bird of prey (*Falconiformes* sp.).[23] It appears that the rodent, reptile, hare, and hedgehog remains are recent and intrusive (Kolska-Horwitz 1986–87).

Examination of the mammalian remains pointed out certain phenomena. On 25 bones (3% of the diagnostic material), cut marks were present that showed signs of butchering, skinning, antler and horn removal, and body dismemberment. However, these cannot be proof for conducting the activities on site. A total of 128 bones (4% of the sample) were found burnt, the largest number of which (57 bones; 44%) were found in the main structure, where most of the activities took place. At least seven caprovines were identified as juveniles, while most of the cattle bones belonged to adult animals. A large number of bones (10% of the diagnostic sample), the majority of which were found in the main structure, were identified as belonging to fallow deer. Unlike other domestic assemblages, no bones of equids, pigs,

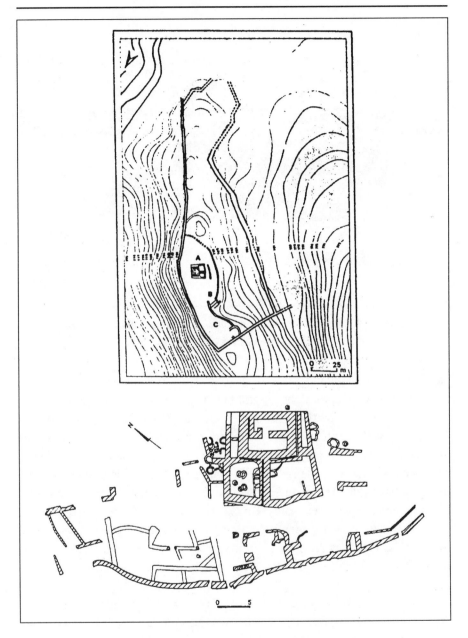

Figure 8.3. *Mt. Ebal, general plan of the site (top)
and the nucleus of the site (bottom).* (Zertal 1986–87:fig. 23.)

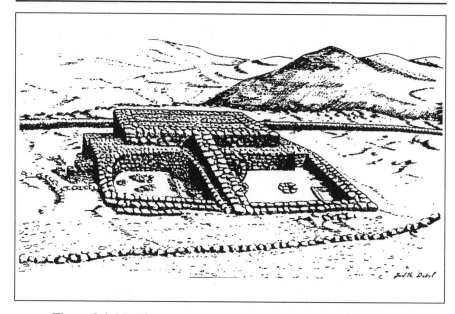

Figure 8.4. *Mt. Ebal, proposed reconstruction of the main structure.*
(Zertal 1986–87:fig 20.)

carnivores (domestic and wild), or gazelles were found at the site. As indicated by Kolska-Horwitz, the assemblage represents a narrow range of activities in function and time. Animals prohibited from consumption are absent, suggesting conformity with biblical prescriptions. Nevertheless, the presence of a large number of fallow deer bones, an animal unacceptable for sacrifice, especially in the area of the main structure, tends to detract from the hypothesis that the structure was an Israelite altar (Kolska-Horwitz 1986-87:187).[24] In a more recent study, Kolska-Horwitz has demonstrated that it is possible to determine the identification of a site as cultic by the examination of bone assemblages (Kolska-Horwitz 1996). She based her study on the comparison of secular and cultic sites closely located temporally and geographically.

    One of the better-known Israelite cult places is the one uncovered at Dan, in northern Galilee on the border with present-day Lebanon.[25] This site was consecrated at the end of the tenth century B.C.E. by Jeroboam I after the cessation of the North (Israel) from the South (Judah) (1 Kgs 12:28–30).

Long-term excavations revealed here a magnificent sacred precinct in which remains of cultic paraphernalia and sacrifices were uncovered (Biran 1994:159–233). Among the last remains from the period preceding Ahab's time, "the finds included a complete bar-handled bowl, with the sign of a trident on its base and in it fragments of sheep, goat and gazelle bones" (Biran 1978:269–70).[26] Many other bones were collected, all of small ruminants. "Of interest is the fact that most of the bones are those on which there would have been very little meat" (Biran 1978:270). More animal bones, this time burnt, were found inside a jar near an altar dated to the time of Jeroboam II (eighth century B.C.E.) (Biran 1994:195).

Zooarchaeological analysis of the material from the Iron Age cult precinct concentrated on material coming from two rooms dated to two periods, one initiated by Ahab (873–851 B.C.E.) and the other by Jeroboam II (785–745 B.C.E.).[27] As Wapnish and Hesse observe, not all processes associated with a cult center can leave traces, especially if the place is in daily use and would be regularly cleaned. Since not all sacrifices are scheduled, the remains uncovered at the site would belong to events that were close in time and space to the end of the site use. The samples from both periods show that "the bulk of the remains are from sheep, goat and cattle, all permissible for sacrifice under Israelite rules, and characteristic of Canaanite sacrifice as well" (Wapnish and Hesse 1991:46). The majority of the bones belong to caprids, while cattle make up 18–29% and deer, 3–9%.[28] A large number of sheep and goats were slaughtered young—25% (on the basis of fusion proportion) were less than one year old; 25–44% (based on dentition) died at six months to one year of age.[29] A similar picture is seen with cattle, of which young animals are also well represented—23% were between one and a half and two or three years old. The same number belongs to animals between two or three and three and a half years old.[30] Other evidence, including a high representation of toes, indicates that collection of skins was actively pursued at the cult precinct (Wapnish and Hesse 1991:35–36). Zooarchaeological analysis suggests that the bones reflect "that slice of sacrificial activity associated with the processing of skins, the consumption of ritual meals, and the storage of remains of burnt offerings" (Wapnish and Hesse 1991:47).

Another Iron Age II (tenth century B.C.E.) sanctuary site that yielded a large number of caprovine (52) and bovid (21) bones is Lachish, a major city

Figure 8.5. *Miniature bronze lion from Arad.* (Herzog 1984:fig 20.)

in the Judean Shephelah. Interestingly, one distal end of a left humerus of a goose (*Anser anser domesticus*) was also found in the sanctuary. It seems that this was a domestic goose (Lernau 1975). Twenty of the caprovine bones can be identified as follows: Seven belong to infants and juveniles (M.N.I. 5);[31] thirteen to subadults and adults (M.N.I. 3). Eight bones belong to goats (M.N.I. 4), two to sheep (M.N.I. 1), and ten to goats or sheep (M.N.I. 3). Of the twenty-one bovid bones, sixteen can be identified as follows: Four belong to juveniles (M.N.I. 1), and eleven to subadults and adults (M.N.I. 2) (Lernau 1975:90). Comparison of the M.N.I. counts agrees with the biblical prescription that calls for the sacrifice of mostly young caprovines.

CARNIVORES IN THE CULT

While no mention is made of the sacrifice or worship of carnivores, skeletal remains at certain sites suggest that in some places the lion was involved in the cult as a worshipped or sacrificed animal.

On a floor of a structure in pre-Israelite Jaffa, possibly a temple of the Late Bronze Age (end of the thirteenth–beginning of the twelfth centuries B.C.E.), a lion skull was found, with half a scarab seal bearing the name of Queen Tiy, wife of Amenhotep III, near the skull's teeth. This building might have been a temple in which a lion cult was practiced (Kaplan and Ritter-Kaplan 1993). Other sites where lion bones were found are Dan and Tel Miqne-Ekron (Wapnish and Hesse 1991:47). At Miqne, several lion bones were found in Field I, in association with Iron Age I pottery in an Iron Age II building the nature of which is still to be determined (Wapnish and Hesse 1991:47). In addition, the same site yielded an amulet depicting a lion. Lion bones (and a bear bone) were found in the Altar Complex at Dan (Wapnish and Hesse 1991:47–48), where on the floor, under the stones of an altar from the time of Jeroboam II, the head of a bronze scepter was disovered. It is 9 cm high and 3.7 cm wide, with four badly corroded figures, possibly depicting lion heads jutting out around a hollow center (Biran 1994). The fact that the former name of Dan is Laish (lion,), the traditions about Samson the Danite and the lion (Judg 14), biblical references to the tribe of Dan as a lion (Deut 33:22), and the material culture presented here all "reinforce the symbolic association of lions and the cult in the Dan sanctuary and by extension within the northern Israelite tradition" (Wapnish and Hesse 1991:48).[32]

It appears that the lion was a cultic element not only in the north, because in Arad, next to the sacrificial altar of the last phase, excavators found a small bronze lion figurine (Herzog et al. 1984:fig. 20; 1987:32) (Fig. 8.5). The association of the tribe of Judah, which supposedly occupied this region, with the lion (Gen 49:9) shows the strong relationship of this area to that animal.

∼

From all of the available evidence, it appears that following biblical instructions, domestic ruminants were the major component of ancient Israel's sacrifices. Birds, especially doves and turtledoves, were also an important sacrificial element, but whether they were raised domestically or caught in the wild is still hard to determine. Based on archaeological evidence, other elements not mentioned in the prescriptions were possibly part of the cultic practices. These include sacrifices of wild ungulates, especially deer and gazelle. Furthermore, the lion, which is a carnivore, figured to a certain degree in the cultic tradition, as evidenced by skeletal remains, artistic depictions in different media, and biblical allusions to tribes and individuals.

Notes

1. Prehistoric and/or historic cult places can be identified not only by their architecture and material culture, but also by several features of the bone assemblage at the site, including the specific animals and body parts represented, age and sex breakdown, bone modifications, and more (Kolska-Horwitz 1996).

2. Examples of this are too many to survey in detail; therefore, from Canaan I would like to cite only the one from Ashkelon, which is relatively close in time and space to the Israelite period (Stager 1991:3, 6–7). In Egypt, the bull represented the god Apis (Pritchard, 1969b:570). It should be mentioned that the cow was also used in Near Eastern iconography to depict godesses, e.g., the Egyptian Hathor. Earlier cultures also revered the bull. For bulls in Crete and bull leaping in Minoan art, see Davidson, 1962:267, 268–9; for calfs in Crete on sarcophagus paintings from Hagia Triada, see Davidson, 1962:260–1.

3. See, for example, the relief from Khorsabad of a man bringing an offering of a goat or ibex (Frankfort 1954:pl. 97), the relief from Carchemish depicting a line of offering bearers, each carrying an ungulate on his shoulder (Pritchard 1969b:fig. 618), or the relief from Karatepe of a man (warrior) carrying on his shoulder what seems to be a calf (Pritchard 1969b:fig. 790).

4. For a good treatment of animal sacrifices, see Wapnish and Hesse 1991, especially pp. 36–47. For a list of the animals, see materials of sacrifices and offerings in Gaster 1962:150–1, 155–6. See schedule of sacrifices in Gaster 1962:150–1.

5. Only freewill offering can be "overgrown or stunted" (Lev 22:23).

6. A donkey, however, could be redeemed with a lamb (Exod 34:20).

7. The prescription of giving the priests the foreleg (*zĕrôʿă*), which is considered the better part, makes doubtful Kenyon's conclusion that the remains in Cave I in Jerusalem are connected with a sanctuary and with Josiah's reforms. While the bones are divided equally between large and small cattle "five times as many bones were found to come from hind limbs than from fore limbs" (Lernau 1995:203).

8. Milgrom describes the nature of this sacrifice in detail (1983).

9. The Hebrew term *ʿărûpâh* stems from the root *ʿRP*, "to break the neck."

10. Examples of events when large numbers of animals were sacrificed are the bringing of the Ark to Jerusalem (2 Sam. 6:13), the coronation of Adonijah (1 Kgs 9, 19, 25), and the inauguration of the Solomonic Temple (1 Kgs 8:5).

11. Two major agricultural events take place in Syria-Palestine at this time of year, the beginning of cereal harvesting (Borowski 1987:31–44) and the beginning of the *raḥil*, transhumance, (see "Herding," chapter 2). Before the combination, each of these feasts celebrated these events separately. The combination itself, however, could have occurred earlier in the history of Israel, but it found expression in the written record only later.

12. Since an animal can be internally sick without external symptoms, selecting it on the tenth of the month and putting it in a quarantine until the fourteenth secures its health.

13. The reference to houses (Exod 12:7, 22) can be an allusion to a tent, in Arabic *bit śa'ar* 'house (made) of hair'.

14. On the *yaḥmûr* as the fallow deer, see Wapnish and Hesse 1991:38–9.

15. The actual term, *ṣārā'at*, refers to some kind of skin disease. When used with inanimate objects, it probably refers to mold or mildew.

16. This is also based on Leviticus 1:14.

17. Although some think that the call is for a young pigeon, the only other place this term is used is in Deuteronomy 32:11: "As an eagle watches over its nest, hovers above its young (*gôzālāyw*), spreads its pinions and takes them up, and bears them on its wings . . . ."

18. A convenient presentation of animal sacrifices is provided in Wapnish and Hesse 1991:table 14.

19. For Mount Ebal as a cult center see below.

20. The case of Jephthah having to sacrifice his daughter (Judg 11:31, 39) is obviously unique.

21. The precise dating of the termination of the temple at Arad is still under discussion. The most recent suggestion for the terminating date is the end of the eighth century B.C.E. (Herzog, personal communication).

22. This identification of the site as a cult center, and specifically as the altar built by Joshua, has been challenged by several archaeologists (for example Kempinski 1986).

23. Other objects found included a piece of eggshell possibly belonging to the grey lag-goose (*Anser anser*), a fragment of a Mediterranean marine shell (*Glycymerys violacescens*), an unidentified fish bone, and land snails (*Levantina caesarina*). The latter are most likely recent and intrusive (Kolska-Horwitz 1986–87:177).

24. For identifying fallow deer with *yaḥmûr*, see above n. 14.

25. Dan is one of the sites studied by Kolska-Horwitz to determine whether a bone assemblage can be used to identify a cult site (see Kolska-Horwitz 1996). The other sites in the study are Horvat Uza and Horvat Qitmit.

26. Cultic vessels and a collection of right foreleg bones of goats were found in a domestic structure at Tell Qiri dated to twelfth-eleventh century B.C.E. (Ben-Tor 1980:35,42), showing that such activities could take place in home shrines, too (see Judg. 17). On the other hand, Kenyon's recovery in Cave I in Jerusalem of parts not fit to be cultic (Lernau 1995), raises the question as to their attribution to Josiah's reforms by the excavator.

27. The material from the cult precinct (Area T) was compared with that from an Iron Age domestic quarter (Area M).

28. At Area M, cattle are nearly 50% and deer 18–20% (Wapnish and Hesse 1991:34). Other animals present at the cult precinct in small numbers include donkey, gazelle, fragment of turtle shell, one bone of small mammal, dog, bear, and lion. None of these are associated with food garbage (Wapnish and Hesse 1991:46).

29. This is twice as much as at Area M, where the pattern is characteristic of the marketable offtake from fiber and meat production (Wapnish and Hesse 1991:34).

30. In Area M, 75% of the animals were over 3 1/2 years old (Wapnish and Hesse 1991:35).

31. M.N.I. = Minimum Number of Individuals, suggesting the possible number of individual animals represented in the sample.

32. See also their earlier report on faunal remain from this site (Wapnish, Hesse et al. 1977).

Chapter 9

The Place of Animals
in the Daily Economy

Biblical and extra-biblical records, coupled with artistic representations and osteological data, reveal the important place animals occupied in Near Eastern cultures in general and in the Israelite economy in particular. Agararian or urbanite, settled or pastoral, the Israelites of the Iron Age (ca. 1200–586 B.C.E.) depended very heavily on animals in every aspect of their economy.

The most important role animals played was their contribution to the daily diet. Most animals were raised for their meat and by-products, and herd management needs determined whether and when animals would be culled and either slaughtered locally or sold in the market. Meat-producing economies tended to eliminate unproductive animals at the end of the suckling period (three months old) or at one year of age. Those removed included male goats and sheep not destined to become studs or produce wool and females that were not going to become ewes or produce wool. Similar criteria were used in the selection of large cattle. Likewise, adult animals at the end of their productive life of breeding or of yielding milk, wool, or hair were also staughtered and replaced by younger ones. The guiding principle was to eliminate any animal that might be an economic burden rather than beneficial. Some of the animals destined for elimination, especially the young ones, were used for sacrifices on different occasions. Biblical prescriptions, supported by osteological evidence, show that mostly young animals were used for sacrifices.

Husbandry of sheep, goats, and cattle was practiced, to a large extent, because of the animal's ability to provide numerous by-products. All of these animals were bred and maintained for their capacity to produce milk in quantities which, in addition to being drunk fresh, could be processed and made into several food-stuffs, such as different kinds of cheese, curds, yoghurt, and other dairy products. These were consumed by the producers themselves and, when surplus was available, were sold in the markets. Moreover, sheep produce wool that can be spun into yarn and then woven into textiles for clothes. Goats grow hair that can be made into ropes and thick cloth for sacks and tents. Large numbers of loom weights, mostly of sun-dried clay, and other tools associated with weaving, such as bone spatulas, spindles, and whorls, indicate that these fibers were actually processed for textiles. Furthermore, goat skins can be made into containers; cow skins can be processed for use as clothing, belts, and shoes, and horns and bones can be utilized in many ways, such as handles for tools, buttons, inlay pieces, or musical instruments. Archaeological data and artistic representations indeed attest that these animal parts were continuously being used. Even dung had its uses; it was collected by the herders and then either spread as organic fertilizer in fields and gardens or sold to others. Although this is difficult to verify archaeologically, there are biblical references alluding to such use.

Husbanding ruminants was thus multifaceted, and every indication supports the notion that the Israelites were very heavily involved in herding these animals, not only in the period of the Settlement (ca. 1200–1000 B.C.E.), but through the First Temple period until its destruction in 587/6 B.C.E. Nomads or transhumants with large herds tended them as a full-time occupation, while others maintained only individual animals for home consumption. The dog, which was probably the earliest animal to be domesticated, was helpful in herding. Not all dogs were employed in herding, however; some were raised as pets or hunting dogs, and others were pariahs roaming the village and city streets or the countryside. Another domestic animal was the cat, however not much is known about it and its place in Ancient Israel.

A by-product of large cattle was muscle power. Individual animals were used as draft animals for pulling wagons, carts, plows, and other agricultural implements. In the Iron Age, oxen and cows were the major source of energy for draft, and they were harnessed daily in the service of the Israelites and their neighbors. They provided most of the power used on the farm and on the road. Joining them, mostly on the road, were the donkey, mule, and camel.

The donkey was the earliest domestic animal of burden. It was first used for packing and draft, but later it was replaced by cattle and was used mostly as a pack animal in all types of terrain. The donkey, which was also utilized for riding, could carry heavy loads on narrow paths in places where no wheeled contraption could pass without extensive investment in road construction. Another equid performing a variety of chores was the mule, a hybrid of donkey with horse. While the mule was used for physical work, it was also considered very valuable and fit to be ridden by kings.

The camel, which as a common domestic animal arrived in Canaan at the end of the Late Bronze Age, filled a unique niche. Its ability to withstand harsh arid conditions made it most suitable for crossing deserts, and it was thus utilized as a pack animal in the caravan trade that crisscrossed the Near East. Moreover, being a fast-running animal, the camel was used also in the military by the camelry. This use was developed by certain ethnic groups, mostly those originating in the Arabian Peninsula. Camel dependency was also fostered by its ability to produce wool, milk, and meat. Zooarchaeological samples suggest that the camel did not enjoy wide distribution in Eretz Yisrael since other animals, such as the donkey and cattle, were more suitable for local needs. A special case is that of Tell Jemmeh in the northwestern Negev, where large numbers of camels were assembled during the Assyrian and Persian periods because of special military and commercial needs.

Another animal with a special niche was the horse. Unlike its modern descendants, in antiquity the horse was reserved for military service and for sport. With the development of the two-wheeled war chariot in the Middle Bronze Age II, the horse replaced the donkey as the energy source for pulling four-wheeled war wagons, and it became the main energy source for pulling these newly developed vehicles, which were used on the battlefield and for the royal sport of hunting. Horseback riding also started at this time, possibly even earlier. Initially, military horseback riders were used for scouting; the use of horses in the cavalry was a much later development that was perfected by the Assyrians. The latter acquired the ability to shoot arrows from a galloping horse controlled by the rider rather than by a companion riding alongside on another horse or together with an archer on the same horse.

Zooarchaeologists suggest that all these mammals were first domesticated for their meat and/or milk, and only later were uses for their other by-products found. Only for one animal, the pig, which was domesticated quite early for its meat, did no other use ever develop. Throughout its domestic history, even to this day, the pig remained a cheap and convenient meat source, but certain cultural traditions have prohibited its consumption. Recently, the historical utilization of the pig has been studied extensively, but no decisive reason has been found for its rejection by the Israelites, a practice that is reflected both in the biblical prohibitions and the osteological record. The latter indicates very clearly that the consumption of pig meat at Israelite sites was either minimal or nil.

The rich inventory of animals, especially mammals, in Syria-Palestine and the surrounding area, and the availability of their by-products, was recognized by the leading powers in the region, who drew on the area as a copious source to fulfill their economic needs. Much of the evidence concerning these animals is available through Egyptian and Mesopotamian written records and artistic representations in different media.

While the mammals discussed above were domestic, most of the birds were wild, though some were kept under controlled conditions. Throughout the ancient Near East, wild birds such as quail, partridge, and water fowl were caught with nets, traps, or bows and arrows and used as a cheap source of meat. While some song birds were kept in cages as pets, other birds, possibly certain kinds of water fowl, were caged and fattened for special occasions. Numerous biblical references mention doves and turtledoves as birds fit for sacrifice and for consumption. Their fitness for the altar and the large numbers that would have been required for cultic purposes strongly suggest that these birds were kept under controlled domestic conditions. Recently discovered osteological and glyptic evidence indicates that, some time during the Iron Age, the domestic chicken arrived to Eretz Yisrael and became one of the barnyard animals. This arrival probably added eggs to the diet, which until then were found and collected only in the wild.

Insects also contributed to the diet in several ways. Since prehistoric times, bees' honey was collected in the wild, as evidenced by paintings from a cave at Bicorp, in the mountains of Valencia in southern Spain. Domesticated bees were already known in Egypt during the Old Kingdom, and there is sufficient written evidence to indicate that honey had been produced domestically in Syria-Palestine, at least since the Late Bronze Age. Some biblical references strongly suggest that the Israelites were also engaged in apiculture and honey production. Honey was eaten raw or used to sweeten other foods, and it was also used as an ingredient in medicines. Wax, the by-product of honey, was used for several purposes the most common of which was the casting of objects in the 'lost wax' method.

Certain insects, such as the locust, were considered delicacies and were eaten when available. Biblical proscriptions permit the consumption of insects belonging to the grasshopper family but there are no references describing such an event. Assyrian reliefs, however, depict servants bringing skewered grasshoppers and garlands of locust to royal feasts.

Another contribution of insects was to provide the dyes kermes (scarlet) and cochineal (crimson). The source for these dyes was a small insect living on the kermes oak, which is native to and common in the Near East. Dyed thread was used in Israelite cultic rituals and must have had some symbolic significance since it was assigned as a marker for the house of Rahab, which helped in the capture of Jericho (Jos 2:18-21).

Yarn and garment dyeing was an important occupation, but not all dyeing was done using the kermes scale insect. A very important and expensive dye source was molluscs of the murex family, which were used for the production of violet and purple dyes. How many Israelites were engaged in murex dyeing is hard to tell. Biblical references mention the need for dyed yarn in certain cultic practices, and a few passages suggest that Israelites learned the dyeing secrets from the Phoenicians. Obviously, those engaged in this craft were settled by the sea shore and thus, might have also been engaged in sailing and maritime trade.

Living by the sea, by river banks, or by the inner lakes led some Israelites to be involved in fishing. Fishing was carried out in boats with different types of nets. Fish-bone remains indicate that processed fish reached landlocked settlements in the hill country, the Negev, and other regions. Some of the fish that found their way to the markets came from distant places as far away as Egypt. These fish were evidently caught, processed, and transported by people other than Israelites. However, that the Israelites had developed a taste for fish is known because fish remains are recovered in almost every excavated site of that period.

Wild animals were also an important element of the faunal inventory. Some of the animals captured locally or abroad were tamed mostly for display purposes, while other were hunted to be used in multiple ways. Deer, gazelle, and other ungulates were primarily hunted for their meat. Their hide, bones, and antlers were also used in ways similar to those of domestic ruminants, as their osteological remains in archaeological excavations testify. For sport, wild carnivores were hunted, and their pelts were worn in religious ceremonies. The skins were also used to cover objects such as the sound-box of musical instruments, and possibly as clothing. Certain skeletal elements, especially of lions, found in Israelite and non-Israelite cult centers suggest that the rituals performed in these sites involved this animal.

Osteological remains and artistic representations show that large wild animals such as the elephant and hippopotamus existed during historical periods in the Near East. While they were never domesticated, they were hunted, especially for their tusks, which were used for carving different kinds of objects including handles, figurines, small boxes, cosmetic implements, inlay pieces, and more. While most people of the biblical period in Eretz Yisrael did not have a chance to see these animals, quite a few saw examples of their products and must have been aware of their existence.

The dependency of Ancient Near Eastern cultures on animals in general, and that of the Israelites in particular, is apparent from the evidence. However, more is needed to be done in this area, and further studies of the relationship between people and animals will promote better understanding of the historical processes that molded this region that has been so influential in shaping our own culture.

References Cited

Abu-Rbei'a, I.
    1990 Goats and Sheep in the Life of the Bedouin. In The Herd and Its
        Products, edited by A. Navon, pp. 9–18 (in Hebrew). Ha-Mador li-Yediat
        ha-Aretz, Tel Aviv.

Aharoni, Y.
    1967a Excavations at Tel Arad: Preliminary Report on the Second Season,
        1963. Israel Exploration Journal 17: 233–49.
    1967b Notes and News: Arad. Israel Exploration Journal 17: 270–2.

Albright, W. F.
    1949 The Archaeology of Palestine. Penguin Books, Harmondsworth,
        Middlesex.

Amiran, R.
    1963 The Ancient Pottery of Eretz Yisrael, from its Beginnings in the
        Neolithic Period to the End of the First Temple (in Hebrew). Bailik
        Institute & Israel Exploration Society, Jerusalem.

Anbar, R.
    1970 Birds of Israel. Yavneh, Tel Aviv.

Angress, S.
    1960 Appendix: The Pig Skeleton from Area B. In Hazor ɪɪ, edited by Y.
        Yadin, Y. Aharoni, R. Amiran, T. Dothan, I. Danayevsky, and J. Perrot,
        pp. 166–174 and pl. ᴄʟxxɪɪɪ. Magnes Press, Jerusalem.

Anthony, D.
    1984   Man and Animals: Living, Working and Changing Together.
           University Museum, University of Pennsylvania, Philadelphia.

Anthony, D. W.
    1991   The Domestication of the Horse. In Equids in the Ancient World,
           edited by R. H. Meadow and H.-P. Uerpmann, pp. 250–77. Dr. Ludwig
           Reichert Verlag, Wiesbaden.

Ariel, D. T.
    1990   Worked Bone and Ivory. In Excavations at the City of David 1978–
           1985, edited by D. T. Ariel, pp. 119–44. Institute of Archaeology, Hebrew
           University of Jerusalem, Jerusalem.

Arter, S.
    1995   Preliminary Report on the Faunal Material from Field IV 1993
           Excavations at Tell Halif. Lahav Research Project, Smithsonian
           Institution, Washington, D.C.

Aufrecht, W. E.
    1995   A Phoenician Seal. In Solving Riddles and Untying Knots: Biblical,
           Epigraphic, and Semitic Studies in Honor of Jonas C. Greenfield, edited
           by Z. Zevit, S. Gitin, and M. Sokoloff, pp. 385–7. Eisenbrauns, Winona
           Lake, Indiana.

Avigad, N.
    1966   A Hebrew Seal with a Family Emblem. Israel Exploration Journal 16:
           50–3, pl. 4c.

Avitsur, S.
    1972   Daily Life in Iretz [sic] Israel in the XIX Century. Am ha-Sepher,
           Tel Aviv.
    1976 Man and His Work: Historical Atlas of Tools and Workshops in the
           Holy Land. Carta and Israel Exploration Society, Jerusalem.

Aynard, J. M.
    1972   Animals in Mesopotamia. In Animals in Archaeology, edited by A. H.
           Brodrick, pp. 42–68. Barrie and Jenkins, London.

Baily, I.
    1978   The Discussion Concerning the Goat: Four Documents. In Papers on
           the Bedouin No. 9, edited by I. Baily, pp. 39–47 (in Hebrew). Sede Boker.

Barber, E. J. W.
    1991 Prehistoric Textiles: The Development of Cloth in the Neolithic and
        Bronze Ages. Princeton University Press, Princeton.

Bate, D. M. A.
    1938 Animal Remains. In Megiddo Tombs, edited by P. L. O. Guy, pp. 209–
        13. University of Chicago Press, Chicago.
    1953 The Animal Bones. In Lachish III: The Iron Age, edited by O. Tufnell,
        pp. 410–1. Oxford University Press, London.

Ben-Tor, A.
    1980 The Regional Study: A New Approach to Archaeological
        Investigation. Biblical Archaeology Review 6(2): 30–48.

Ben-Tuvia, A.
    1982 On the Taxonomic Classification of the Fish Weight. Atiqot (English
        Series) 15: 100.

Bennett, W. J. J., and J. H. Schwartz
    1989 Faunal Remains. In Tell el-Hesi, The Persian Period (Stratum V),
        edited by W. J. J. Bennett and J. A. Blakely, pp. 257–63. Eisenbrauns,
        Winona Lake, Indiana.

Bietak, M., ed.
    1990 Ägypten und Levante. Austrian Academy of Science, Vienna.

Biran, A.
    1978 Notes and News: Tel Dan. Israel Exploration Journal 28: 268–71.
    1994 Biblical Dan. Israel Exploration Society and Hebrew Union College-
        Jewish Institute of Religion, Jerusalem.

Bodenheimer, F. S.
    1935 Animal Life in Palestine. L. Mayer, Jerusalem.
    1960 Animal and Man in Bible Lands. E. J. Brill, Leiden.

Bökönyi, S.
    1989 Definition of Animal Domestication. In The Walking Larder: Patterns
        of Domestication, Pastoralism, and Predation, edited by J. Clutton-Brock,
        pp. 22–7. Unwin Hyman, London.

Bokser, B. M.
> 1992 Unleavened Bread and Passover, Feasts of. In The Anchor Bible Dictionary, edited by D. N. Freedman, pp. 755–65. Doubleday, New York.

Borowski, O.
> 1983 The Identity of the Biblical Ṣirʿâ. In The Word of the Lord Shall Go Forth: Essays in Honor of D. N. Freedman in Celebration of His Sixtieth Birthday, edited by C. L. Meyers and M. O'Connor, pp. 315–91. Eisenbrauns, Winona Lake, Indiana.
> 1987 Agriculture in Iron Age Israel. Eisenbrauns, Winona Lake, Indiana.

Bridgeman, J.
> 1987 Purple Dye in Late Antiquity and Byzantium. In The Royal Purple and the Biblical Blue: *Argaman* and *Tekhelet*, edited by E. Spanier, pp. 159–65. Keter, Jerusalem.

Brothwell, D., and P. Brothwell
> 1969 Food in Antiquity: A Survey of the Diet of Early Peoples. Frederick A. Praeger, New York.

Brown, F., S. R. Driver, and C. A. Briggs
> 1906 Hebrew and English Lexicon of the Old Testament. Houghton, Mifflin and Co., Cambridge, Massachusetts.

Buhl, F.
> 1910 Wilhelm Gesenius' Hebräisches und Aramäisches Handwürterbuch über das Alte Testament. Verlag von F. C. W. Vogel, Leipzig.

Buitenhuis, H.
> 1991 Some Equid Remains from South Turkey, North Syria, and Jordan. In Equids in the Ancient World, edited by R. H. Meadow and H.-P. Uerpmann. Dr. Ludwig Reichert Verlag, Wiesbaden.

Bulliet, R. W.
> 1975 The Camel and the Wheel. Harvard University Press, Cambridge.

Burleigh, R.
> 1986 Chronology of Some Early Domestic Equids in Egypt and Western Asia. In Equids in the Ancient Near East, edited by R. H. Meadow and H.-P. Uerpmann. Dr. Ludwig Reichert Verlag, Wiesbaden.

Burleigh, R., J. Clutton-Brock, and J. Gowlett
    1991   Early Domestic Equids in Egypt and Western Asia: An Additional
        Note. In Equids in the Ancient World, edited by R. H. Meadow and H.-P.
        Uerpmann. Dr. Ludwig Reichert Verlag, Wiesbaden.

Carrington, R.
    1972   Animals in Egypt. In Animals in Archaeology, edited by A. H.
        Brodrick, pp. 69–89. Barrie and Jenkins, London.

Clutton-Brock, J.
    1981   Domesticated Animals from Early Times. University of Texas Press,
        Austin, and British Museum, London.
    1989   Cattle in Ancient North Africa. In The Walking Larder: Patterns of
        Domestication, Pastoralism, and Predation, edited by J. Clutton-Brock,
        pp. 200–6. Unwin Hyman, London.
    1992   Horse Power: A History of the Horse and Donkey in Human Societies.
        Harvard University Press, Cambridge.

Cohen, M. N.
    1977   The Food Crisis in Prehistory: Overpopulation and the Origin of
        Agriculture. Yale University Press, New Haven and London.

Cohen, R., and Y. Yisrael
    1995   On the Road to Edom: Discoveries from 'En Hazeva. Israel Museum,
        Jerusalem.

Cole, S.
    1972   Animals in the New Stone Age. In Animals in Archaeology, edited by
        A. H. Brodrick., pp. 15–41. Barrie and Jenkins, London.

Crane, E.
    1975   History of Honey. Honey: A Comprehensive Survey, edited by E.
        Crane, pp. 439–88. Heinemann, London.
    1983   The Archaeology of Beekeeping. Cornell University Press, Ithaca,
        New York.

Cross, F. M.
    1973   Canaanite Myth and Hebrew Epic. Harvard University Press,
        Cambridge.

Crowfoot, J. W., and G. M. Crowfoot
　　1938　Early Ivories from Samaria. Palestine Exploration Fund, London.

Dalley, S.
　　1984　Mari and Karana. Longman, London and New York.

Davidson, B.
　　1848　The Analytical Hebrew and Chaldee Lexicon. Samuel Bagster and
　　　　　Sons, London.

Davidson, M. B.
　　1962　The Horizon Book of Lost Worlds. American Heritage Publishing,
　　　　　New York.

Davidson, M. B. (editor)
　　1965　The Light of the Past. American Heritage Publishing, New York.

Davies, B.
　　1883　A Compandious and Complete Hebrew and Chaldee Lexicon to the
　　　　　Old Testament. Warren F. Draper, Andover, Massachusetts.

Davis, S. M. J.
　　1987　The Archaeology of Animals. Yale University Press, New Haven and
　　　　　London.

Dayan, T.
　　1993　The Impact of Quarternary Paleoclimatic Change on the Carnivores of
　　　　　Israel. Water, Science, and Technology 27(7–8): 497–504.
　　1994a　Carnivore Diversity in the Late Quarternary of Israel. Quarternary
　　　　　Research 41: 343–9.
　　1994b　Early Domesticated Dogs of the Near East. Journal of Archaeological
　　　　　Science 21(5):633–40.
　　1996　Ecology, Evolution, and Zooarchaeology: The Analysis of Mammalian
　　　　　Remains from Archaeological Sites. Paper delivered in the conference
　　　　　The Practical Impact of Science on Field Archaeology: Jerusalem and Tel
　　　　　Aviv, October 28–29. Organized by the Malcolm H. Wiener Laboratory of
　　　　　the American Schools of Classical Studies at Athens and the W. F.
　　　　　Albright Institute of Archaeological Research, Jerusalem.

Dayan, T., and D. Simberloff
　　1995　Natufian Gazelles: Proto-Domestication Reconsidered. Journal of
　　　　　Archaeological Science 22:671–5.

Dayan, T., D. Simberloff, and E. Tchernov
1990   Feline Canines: Community-Wide Character Displacement among the Small Cats of Israel. The American Naturalist 136(1 July):39–60.

Dayan, T., D. Simberloff, E. Tchernov, and Y. Yom-Tov
1991   Calibrating the Paleothermometer: Climate, Communities, and the Evolution Size. Paleobiology 17(2):189–99.
1992   Canine Carnassials: Character Displacement in the Wolves, Jackals and Foxes of Israel. Biological Journal of the Linnean Society 45:315–31.

Dent, A.
1972   Donkey, The Story of the Ass from East to West. Yale University Press, London.

Edgerton, W. F. and J. A. Wilson
1936   Historical Records of Rameses III: The Texts in Medinet Habu. Vols. I and II. University of Chicago Press, Chicago.

Elat, M.
1977   Economic Relations in the Lands of the Bible c.1000–539 B.C. Mosad Bialik & Israel Exploration Fund, Jerusalem.

Epstein, H.
1985   The Awassi Sheep with Special Reference to the Improved Dairy Type. Food and Agriculture Organization of the United Nations, Rome.

Finkelstein, I.
1986   'Izbet Sartah: An Early Iron Age Site near Rosh Ha'ayin, Israel. BAR International Series 299. British Archaeological Reports, Oxford.

Firouz, L.
1972   The Caspian Miniature Horse of Iran. Edwards Brothers, Miami.

Flannery, K. V.
1973   The Origins of Agriculture. In Annual Review of Anthropology 2:271–310.

Forbes, R. J.
1956   Studies in Ancient Technology. E. J. Brill, Leiden.

Frame, G.
    1995   Rulers of Babylonia: From the Second Dynasty of Isin to the End of
    the Assyrian Domination. University of Toronto Press, Toronto.

Frankfort, H.
    1954   The Art and Architecture of the Ancient Orient. Penguin Books,
    Harmondsworth, Middlesex.

Fraser, H. M.
    1931   Beekeeping in Antiquity. University of London Press, London.

Gaster, T. H.
    1962   Sacrifices and Offerings, OT. In The Interpreter's Dictionary of the
    Bible, edited by G. A. Buttrick, pp. 147–59. Abingdon Press, New York
    and Nashville.

Gilbert, A. S.
    1991   Equid Remains from Godin Tepe, Western Iran: An Interim Summary
    and Interpretation, with Notes on the Introduction of the Horse into
    Southwest Asia. In Equids in the Ancient World, edited by R. H. Meadow
    and H.-P. Uerpmann, 75–110. Dr. Ludwig Reichert Verlag, Wiesbaden.

Ginat, Y.
    1966   Wandering. In The Bedouin, pp. 43–7 (in Hebrew). School for Tourist
    Guidance, Jerusalem.

Grigson, C.
    1993   The Earliest Domestic Horses in the Levant?—New Finds from the
    Fourth Millennium of the Negev. Journal of Archaeological Science
    20(6):645–55.
    1995   Plough and Pasture in the Early Economy of the Southern Levant. In
    The Archaeology of Society in the Holy Land, edited by T. E. Levy, pp.
    245–68. Leicester University Press, London.

Groves, C. P.
    1986   The Taxonomy, Distribution, and Adaptations of Recent Equids. In
    Equids in the Ancient World, edited by R. H. Meadow and H.-P.
    Uerpmann, pp. 11–51. Dr. Ludwig Reichert Verlag, Wiesbaden

Haas, N.
    1971   Anthropological Observations on the Skeletal Remains Found in
    Area D (1962–1963). Atiqot (English Series) 9–10:212–3.

Hadley, J.
1988 Review of *Agriculture in Iron Age Israel* by O. Borowski. In Vetus Testamentum 38 (October): 494–5.

Hakker-Orion, D.
1984 The Role of the Camel in Israel's Early History. In Animals and Archaeology: Early Herders and their Flocks, edited by J. Clutton-Brock and C. Grigson, pp. 207–221. British Archaeological Reports, Oxford.

Hall, H. R.
1928 Babylonian and Assyrian Sculpture in the British Museum. Les Éditions G. Van Oest, Paris and Brussels.

Haran, M.
1985 Temples and Temple-Service in Ancient Israel. Eisenbrauns, Winona Lake, Indiana.

Hellwing, S.
1984 Human Exploitation of Animal Resources in the Early Iron Age Strata at Tel Beer-Sheba. In Beer-Sheba II: The Early Iron Age Settlements, edited by Z. Herzog, pp. 105–15. Tel Aviv University, The Institute of Archaeology and Ramot Publishing, Tel Aviv.
1988-89 Faunal Remains from the Early Bronze and Late Bronze Ages at Tel Kinrot. Tel Aviv 15–16(2):212–20.

Hellwing, S., and Y. Adjeman
1986 Animal Bones. In 'Izbet Sartah: An Early Iron Age Site near Rosh Ha'ayin, Israel, edited by I. Finkelstein, pp. 141–52. BAR International Series 299. British Archaeological Reports, Oxford.

Hellwing, S., and N. Feig
1988 Animal Bones. In Excavations at Tel Michal, Israel, edited by Z. Herzog, G. Rapp Jr., and O. Negbi, pp. 236–247. University of Minnesota Press, Minneapolis, and Tel Aviv University, Tel Aviv.

Herzog, Z.
1984 Beersheba II: The Early Iron Age Settlements. Tel Aviv University, The Institute of Archaeology and Ramot Publishing, Tel Aviv.

Herzog, Z., M. Aharoni, and A. F. Rainey
1987 Arad: An Ancient Israelite Fortress with a Temple to Yahweh. Biblical Archaeology Review 13(2):16–35.

Herzog, Z., M. Aharoni, A. F. Rainey, and S. Moskovitz
1984 Israelite Fortress at Arad. Bulletin of the American Schools of Oriental Research 254:1–34.

Hess, R. S.
1993 Early Israel in Canaan: A Survey of Recent Evidence and Interpretations. Palestine Exploration Quarterly 125(July–December):125–142.

Hesse, B.
1990 Pig Lovers and Pig Haters: Patterns of Palestinian Pork Production. Journal of Ethnobiology 10(2):195–225.

Hesse, B., and P. Wapnish
1985 Animal Bone Archaeology. Taraxacum, Washington.
1994 Can Pig Remains Be Used for Ethnic Diagnosis in the Ancient Near East. Paper delivered in the conference The Archaeology of Israel: Constructing the Past/Interpreting the Present. Lehigh University, Lehigh, Pennsylvania.
1996 Pigs' Feet, Cattle Bones and Birds' Wings. Biblical Archaeology Review 22(1):62.

Hirsch, S.
1933 Sheep and Goats in Palestine. Palestine Economic Society, Tel Aviv.

Holladay, J. S. J.
1986 The Stables in Ancient Israel. In The Archaeology of Jordan and Other Studies, edited by L. T. Geraty and L. G. Herr, 103–65. Andrews University Press, Berrien Springs, Michigan.

Holladay, W. L.
1978 A Concise Hebrew and Aramaic Lexicon of the Old Testament.William B. Eerdmans, Grand Rapids, Michigan.

Holland, T. A.
1977 A Study of Palestinian Iron Age Baked Clay Figurines, with Special Reference to Jerusalem: Cave 1. Levant 9:121–55.

Ilan, S.
1984 Small Cattle. In Eretz Yisrael Landscapes in the 19th Century and Traditional Arab Agriculture, edited by E. Schiller, pp. 65–7. Kardom, Jerusalem.

Ishak, R. R.
1987 Improving Small-Scale Sheep and Goat Cheesemaking in the Near East. In Small Ruminants in the Near East, edited by A. W. Qureshi and H. A. Fitzhugh, pp. 221–36. Food and Agriculture Organization of the United Nations, Rome.

Jones, E. J.
1976 Hives and Honey of Hymettus: Beekeeping in Ancient Greece. Archaeology 29(2):80–91.

Kaplan, J., and H. Ritter-Kaplan
1993 Jaffa. In The New Encyclopedia of Archaeological Excavations in the Holy Land, edited by E. Stern, pp. 655–59. Carta and Israel Exploration Society, Jerusalem.

Kaplan, M.
1989 The Case of the Goat. Teva Ve'eretz 31(4):27–30.

Karmon, N., and E. Spanier
1987 Archaeological Evidence of the Purple Dye Industry from Israel. In The Royal Purple and the Biblical Blue: *Argaman* and *Tekhelet*, edited by E. Spanier, pp. 147–58. Keter, Jerusalem.
1988 Remains of a Purple Dye Industry Found at Tel Shiqmona. Israel Exploration Journal 38(3):184–6.

Kempinski, A.
1986 Joshua's Altar—An Iron Age Watchtower. Biblical Archaeology Review 12:42, 44–9.

Koehler, L., and W. Baumgartner
1953 Lexicon in Veteris Testamenti Libros. Brill, Leiden.

Köhler, I.
1984 The Dromedary in Modern Pastoral Societies and Implications for its Process of Domestication. In Animals and Archaeology: Early Herders and their Flocks, edited by J. Clutton-Brock and C. Grigson, pp. 201–6. British Archaeological Reports, Oxford.

Kolska-Horwitz, L.
1986–87 Faunal Remains from the Early Iron Age Site on Mount Ebal. Tel Aviv 13–14(2):173–89.
1990 Archaeozoological Analysis of Raw Materials Used in the Manufacture of Bone Artifacts. In Excavations at the City of David 1978–1985, edited by D. T. Ariel, pp. 144–5. Institute of Archaeology, Hebrew University of Jerusalem, Jerusalem.
1993 Note 2: Bone Remains. In Ashdod v: Excavation of Area G, edited by M. Dothan and Y. Porath, p. 144. Israel Antiquities Authority, Jerusalem.
1996 The Contribution of Archaeozoology to the Identification of Ritual sites. Paper delivered in the conference The Practical Impact of Science on Field Archaeology: Jerusalem and Tel Aviv, October 28–29. Organized by the Malcolm H. Wiener Laboratory of the American Schools of Classical Studies at Athens and the W. F. Albright Institute of Archaeological Research, Jerusalem.

Kolska-Horwitz, L., and E. Tchernov
1989 Subsistence Patterns in Ancient Jerusalem: A Study of Animal Remains. In Excavations in the South of the Temple Mount: The Ophel of Biblical Jerusalem, edited by E. Mazar and B. Mazar, pp. 144–54. Institute of Archaeology, Hebrew University of Jerusalem, Jerusalem.

Koren, Z.
1996 What is "*tōla'at haššānî*" in Archaeological Textiles? New Chemical Researches. Twenty-second Archaeological Conference in Israel, Tel Aviv University. Israel Exploration Society and Israel Antiquities Authority.

Kozloff, A. P. (editor)
1981 Animals in Ancient Art from the Leo Mildenberg Collection. Cleveland Museum of Art, Cleveland, Ohio, and Indiana University Press, Bloomington.

Krispill, N.
1986 Milk and Its Processing by the Fellahin of H. Susiya. In Cave-Dwelling Fellahin, edited by A. Naron, pp. 26–30 (in Hebrew). Ha-Mador li-Yediat ha-Aretz, Tel Aviv.

Legge, A. J.
1978 Archaeozoology—or Zooarchaeology? In Research Problems in Zooarchaeology, edited by D. R. Brothwell, K. D. Thomas, and J. Clutton-Brock, pp. 129–32. Institute of Archaeology, University of London, London.

Lernau, H.
1975　Animal Remains. In Investigations at Lachish: The Sanctuary and the Residency (Lachish v), edited by Y. Aharoni, pp. 86–103. Gateway Publishers, Tel Aviv.
1986　Fishbones Excavated in Two Late Roman-Byzantine Castella in the Southern Desert of Israel. In Fish and Archaeology, edited by D. C. Brinkhuizen and A. T. Clason, pp. 85–102. BAR International Series 294. British Archaeological Reports, Oxford.
1995　Faunal Remains from Cave I in Jerusalem. In Excavations by K.M. Kenyon in Jerusalem 1961–1967, edited by I. Eshel and K. Prag, pp. 201–8. British School of Archaeology in Jerusalem, Oxford.

Lernau, H., and O. Lernau
1989　Fish Bone Remains. In Excavations in the South of the Temple Mount: The Ophel of Biblical Jerusalem, edited by E. Mazar and B. Mazar, pp. 155–9. Institute of Archaeology, Hebrew University of Jerusalem. Jerusalem.
1992　Fish Remains. In Excavations at the City of David 1978–1985, edited by A. De Groot and D. T. Ariel, pp. 131–48. Institute of Archaeology, Hebrew University of Jerusalem, Jerusalem.

Leteris, M. H. (editor)
1984　The Holy Scriptures of the Old Testament: Hebrew and English. The British and Foreign Bible Society, London.

Levi, S.
1978　*Riḥleh*—Looking for Pasturage. Teva ve-Eretz 20:270–7.

Levy, T. E.
1991　Dogs and Healing. Biblical Archaeology Review 17(6):14–8.

Licht, J. S.
1971　Passover. In Encyclopaedia Biblica, edited by B. Mazar, pp. 514–26 (in Hebrew). Mosad Bialik, Jerusalem.

Lowdermilk, W. C.
1944　Palestine—Land of Promise. Harper and Brothers, New York.

Lydekker, R.
1912　The Horse and Its Relatives. George Allen, London.

Macalister, R. A. S.
    1912   Excavations of Gezer 1902–1905 and 1907–1909. Palestine
           Exploration Fund, London.

Mandelkern, S.
    1967   Veteris Testamenti Concordantiae Hebraicae atque Chaldaicae.
           Shoken, Tel Aviv.

Mayor, A.
    1995   Mad Honey! Archaeology 48(6):32–40.

Mazar, A.
    1982   The 'Bull Site'—An Iron Age I Open Cult Place. Bulletin of the
           American Schools of Oriental Research 247:27–42.
    1990   Archaeology of the Land of the Bible. Doubleday, New York.

Meadow, R. H., and H.-P. Uerpmann, eds.
    1986   Equids in the Ancient Near East. Dr. Ludwig Reichert Verlag,
           Wiesbaden.
    1991   Equids in the Ancient World. Dr. Ludwig Reichert Verlag, Wiesbaden.

Merhav, R. (editor)
    1987   Treasures of the Bible Lands: The Ellie Borowski Collection. Tel Aviv
           Museum and Modan Publishers, Tel Aviv.

Meshel, Z.
    1974   New Data about the "Desert Kites." Tel Aviv 1:129–43.
    1978   Kuntillet 'Ajrud: A Religious Centre from the Time of the Judaean
           Monarchy on the Border of Sinai. The Israel Museum, Jerusalem.
    1993   Teman, Horvat. In The New Encyclopedia of Archaeological
           Excavations in the Holy Land, edited by E. Stern, pp. 1458–1464. Carta
           and Israel Exploration Society, Jerusalem.

Michalowski, K.
    1968   Art of Ancient Egypt. Harry N. Abrams, New York.

Mienes, H. K.
    1987   A Review of the Family Janthinidae (Mollusca, Gastropoda) in
           Connection with the Tekhelet Dye. In The Royal Purple and the Biblical
           Blue: Argaman and Tekhelet, edited by E. Spanier, pp. 197–205. Keter,
           Jerusalem.

1992   Molluscs. In Excavations at the City of David 1978–1985, edited by A. De Groot and D. T. Ariel, pp. 122–30. Institute of Archaeology, Hebrew University of Jerusalem, Jerusalem.

Milgrom, J.
1983   Studies in Cultic Theology and Terminology. E. J. Brill, Leiden.
1991   The Anchor Bible: Leviticus 1–16. Doubleday, New York.

Mohr, E.
1971   The Asiatic Wild Horse: Equus przevalskii Poliakoff, 1881. J. A. Allen, London.

Muzzolini, A.
1983   Les Types de Boeufs Domestiques dans l'Égypte Ancienne. Bulletin de la Sociéte Meridionale de Speleologie et Préhistoire 23:55–75.

Netzer, E.
1992   Domestic Architecture in the Iron Age. In The Architecture of Ancient Israel: From the Prehistoric to the Persian Periods, edited by A. Kempinsky and R. Reich, pp. 193–201. Israel Exploration Society, Jerusalem.

Neufeld, E.
1978   Apiculture in Ancient Palestine (Early and Middle Iron Age) within the Framework of the Ancient Near East. In Ugarit-Forschungen, vol. 10, pp. 219–47. Verlag Butzon and Bercker Kevelaer, Münster.
1980   Insects as Warfare Agents in the Ancient Near East. Orientalia 49:30–57.

Newberry, P. E.
1893, 1894   Beni Hasan. Egypt Exploration Fund, London.

Nissen, H. J.
1988   The Early History of the Ancient Near East 9000–2000 B.C. University of Chicago, Chicago and London.

Nun, M.
1993   Cast Your Net upon the Waters. Biblical Archaeology Review 19 (6):46–56, 70.

Olsen, S. J.
> 1985 Origins of the Domestic Dog: The Fossil Record. University of Arizona Press, Tucson.

Oppenheim, A.
> 1996 DNA—A Means for History Research. Twenty-Second Archaeological Conference in Israel, Tel Aviv University. Israel Exploration Society and Israel Antiquities Authority.

Pardee, D.
> 1978 Letters from Tel Arad. In Ugarit-Forschungen, vol. 10, pp. 289–336. Verlag Butzon and Bercker Kevelaer, Münster.

Parotiz, J. J.
> 1982 Note on Invertebrata. Atiqot (English Series) 15:101.

Postgate, J. N.
> 1986 The Equids of Sumer, Again. In Equids in the Ancient World, edited by R. H. Meadow and H.-P. Uerpmann, pp. 194–213. Dr. Ludwig Reichert Verlag, Wiesbaden.

Pravolotsky, A., and A. Pravolotsky
> 1979 Herd and Grazing in the Sinai Mountains. In Gardening and Herding: Traditional Livelihood Sources of the Jabaliya Tribe, pp. 55–95 (in Hebrew). Tsukey David Field School, Israel.

Pritchard, J. B.
> 1970 The Megiddo Stables: A Reassessment. In Essays in Honor of Nelson Glueck: Near Eastern Archaeology in the Twentieth Century, 268–76. Doubleday and Co., New York.

Pritchard, J. B., editor
> 1958 The Ancient Near East, An Anthology of Texts and Pictures. Princeton University Press, Princeton.
> 1969a Ancient Near Eastern Text Relating to the Old Testament. Princeton University Press, Princeton.
> 1969b The Ancient Near East in Pictures Relating to the Old Testament. Princeton University Press, Princeton.

Ratner, R., and B. Zuckerman
> 1986 'A Kid in Milk'?: New Photographs of KTU 1.23, Line 14. In Hebrew Union College Annual, edited by S. H. Blank, pp. 15–60. Hebrew Union College–Jewish Institute of Religion, Cincinnati.

Rose, R. J., and D. R. Hodgson
1993   Manual of Equine Practice. W. B. Saunders, Philadelphia.

Rosen, B.
1995   Pig Raising in Eretz Yisrael after the Roman Period. Kathedra 78:25–42 (in Hebrew).

Rylaarsdam, J. C.
1962   Passover and Feast of Unleavened Bread. In Interpreter's Dictionary of the Bible, edited by G. A. Buttrick, pp. 663–668. Abingdon Press, New York and Nashville.

Salonen, A.
1976   Jagd und Jagdtiere im alten Mesopotamien. N.p., Helsinki.

Sanderson, I. T.
1955   Living Mammals of the World. Hanover House, Garden City, New Jersey.

Schulman, A. R.
1957   Egyptian Representations of Horsemen and Riding in the New Kingdom. Journal of Near Eastern Studies 16:236–71.

Seger, J. D.
1991   Tel Halif—1989. Excavations and Surveys in Israel 9:67–8.

Seger, J. D., B. Baum, O. Borowski, D. P. Cole, H. Forshey, E. Futato, P. F. Jacobs, M. Laustrup, P. O'Connor Seger, M. Zeder
1990   The Bronze Age Settlements at Tell Halif: Phase II Excavations, 1983-1987. Bulletin of the American Schools Supplements 26:1–32.

Sherratt, A.
1983   The Secondary Exploitation of Animals in the Old World. World Archaeology 15:90–104.

Shkolnik, A.
1977   Physiological Adaptation of Mammals to Desert Life. In The Desert: Past, Present, Future, edited by A. Zohar, pp. 100–12 (in Hebrew). Reshafim, Tel Aviv.

Shulow, A.
1967   Names of Birds in the Bible. In Birds of Israel, 3d ed., edited by P. Arnold, pp. 106–107. Shalit Publishers, Haifa.

Simpson, G. G.
    1951 Horses: The Story of the Horse Family in the Modern World and
    through Sixty Million Years of History. Oxford University Press, New
    York.

Spanier, E. (editor)
    1987 The Royal Purple and the Biblical Blue: 'Argāmān and Tĕklet. Keter,
    Jerusalem.

Spanier, E., and N. Karmon
    1987 Muricid Snails and the Ancient Dye Industries. In The Royal Purple
    and the Biblical Blue: 'Argāmān and Tĕklet, edited by E. Spanier, pp.
    179–96. Keter, Jerusalem.

Stager, L. E.
    1991 Ashkelon Discovered: From Canaanites and Philistines to Romans and
    Moslems. Biblical Archaeology Society, Washington, D.C.

Steindorff, G.
    1946 Catalogue of the Egyptian Sculpture in the Walters Art Gallery. The
    Trustees/Walters Art Gallery, Baltimore.

Stern, E., and I. Sharon
    1987 Tel Dor, 1986. Israel Exploration Journal 37(4):201–11.

Stevenson, T. B., and B. Hesse
    1990 "Domestication" of Hyrax (Procavia capensis), in Yemen. Journal of
    Ethnobiology 10(1):23–32.

Stieglitz, R. R.
    1994 The Minoan Origin of Tyrian Purple. Biblical Archaeologist 57:46–54.

Suggs, M. J., K. D. Sakenfeld, and J. R. Mueller (editors)
    1992 The Oxford Study Bible: Revised English Bible with the Apocrypha.
    Oxford University Press, New York.

Taylor, J. G.
    1993 Yahweh and the Sun: Biblical and Archaeological Evidence for Sun
    Worship in Ancient Israel. JSOT Press, Sheffield, England.

Tchernov, E., T. Dayan, and Y. Yom-Tov
1986/87 The Paleogeography of *Gazella Gazella* and *Gazella Dorcas* during the Holocene of the Southern Levant. Israel Journal of Zoology 34:51–9.

Uerpmann, H.-P.
1987 The Ancient Distribution of Ungulate Mammals in the Middle East. Dr. Ludwig Reichert Verlag, Wiesbaden.

Ussishkin, D.
1982 The Conquest of Lachish by Sennacherib. Tel Aviv University/Institute of Archaeology, Tel Aviv.

Van Buren, E. D.
1930 Clay Figurines of Babylonia and Assyria. Yale University Press, New Haven, and Oxford University Press, London.
1945 Symbols of the Gods in Mesopotamian Art. Pontifical Biblical Institute, Rome.

von Soden, W.
1994 The Ancient Near East: An Introduction to the Study of the Ancient Near East. William B. Eerdmans, Grand Rapids, Michigan.

Wapnish, P.
1984 The Dromedary and Bactrian Camel in Levantine Historical Settings: The Evidence from Tell Jemmeh. In Animals and Archaeology: Early Herders and their Flocks, edited by J. Clutton-Brock and C. Grigson, pp. 171–200. British Archaeological Reports, Oxford.
1991 Beauty and Utility in Bone: New Light on Bone Crafting. In Ashkelon Discovered: From Canaanites and Philistines to Romans and Moslems. Biblical Archaeology Society, Washington, D.C.
1993 Archaeozoology: The Integration of Faunal Data with Biblical Archaeology. In Biblical Archaeology Today, 1990, edited by A. Biran and J. Aviram, pp. 426–42. Israel Exploration Society and Israel Academy of Sciences and Humanities, Jerusalem.
1996 Is *seni ana la mani* an Accurate Description or a Royal Boast? In Retrieving the Past: Essays on Archaeological Research and Methodology in Honor of Gus W. Van Beek, edited by J. D. Seger, pp. 283–94. Cobb Institute of Archaeology and Eisenbrauns, Winona Lake, Indiana.

Wapnish, P., and B. Hesse
1991 Faunal Remains from Tel Dan: Perspectives on Animal Production at a Village, Urban and Ritual Center. ArchaeoZoologia 4(2):9–86.
1993 Pampered Pooches or Plain Pariahs? The Ashkelon Dog Burials. Biblical Archaeologist 56:55–80.

Wapnish, P., B. Hesse, and A. Ogilvy
1977 The 1974 Collection of Faunal Remains from Tell Dan. Bulletin of the American Schools of Oriental Research 227:35–62.

Watkins, T.
1989 The Beginning of Warfare. In Warfare in the Ancient World, edited by J. Hackett, pp. 15–35. Facts on File, New York.

West, B., and B.-X. Zhou
1988 Did Chickens Go North? New Evidence for Domestication. Journal of Archaeological Science 15:515–33.

Wheeler, A.
1978 Problems of Identification and Interpretation of Archaeological Fish Remains. In Research Problems in Zooarchaeology, edited by D. R. Brothwell, K. D. Thomas, and J. Clutton-Brock, pp. 69–75. Institute of Archaeology, University of London, London.

Wilson, L. M.
1933 Ancient Textiles from Egypt in the University of Michigan Collection. University of Michigan Press, Ann Arbor.

Wiseman, D. J.
1989 The Assyrians. In Warfare in the Ancient World, edited by J. Hackett, pp. 36–53. Facts on File, New York.

Yadin, Y.
1963 Warfare in the Lands of the Bible. International Publishing Co., Ramat Gan, Israel.
1975 The Meggido Stables. In Eretz Israel: Nelson Glueck Memorial Volume, edited by B. Mazar, 57–62. Israel Exploration Fund and Hebrew Union College.

Yagil, R.
    1993   From Its Blood to Its Hump, the Camel Adapts to the Desert. Natural
    History 102:30–3.

Yalçin, B. C.
    1986   Sheep and Goats in Turkey. Food and Agriculture Organization of the
    United Nations, Rome.

Zarins, J.
    1986   Equids Associated with Human Burials in Third Millennium B.C.
    Mesopotamia: Two Complementary Facets. In Equids in the Ancient
    World, edited by R. H. Meadow and H.-P. Uerpmann, pp. 164–193. Dr.
    Ludwig Reichert Verlag, Wiesbaden.

Zeder, M. A.
    1996   The Role of Pigs in Near Eastern Subsistence: A View from the
    Southern Levant. In Retrieving the Past: Essays on Archaeological
    Research and Methodology in Honor of Gus W. Van Beek, edited by J. D.
    Seger, pp. 295–310. Cobb Institute of Archaeology and Eisenbrauns,
    Winona Lake, Indiana.

Zertal, A.
    1986–87 An Early Iron Age Cultic Site on Mount Ebal: Excavation Seasons
    1982–1987. Tel Aviv 13-14(2):105–65.

Zorn, J. R.
    1993   Nasbeh, Tell en-. In The New Encyclopedia of Archaeological
    Excavations in the Holy Land, edited by E. Stern, pp. 1098–102. Carta
    and Israel Exploration Fund, Jerusalem.

Subject Index

Archaeological Sites

Authors Cited

271

Terms in Hebrew and Other Semitic Languages

Alphabetized according to Hebrew consonantal order,
including final *hē* and the *matres lectionis*; vowels are ignored.

'ēbûs	trough	129
'abbîr	bull, mighty	213
'ăbîr ya'aqob	mighty one of Jacob	78, 211
'ăbîr yiśrā'ēl	mighty one of Israel	78, 211
'ôpān	wheel	89
'ayyâh	kite	149
'ayyāl	roe deer	186, 207, 218
'ayyālâh	doe	186
'ayil	ram	68, 215, 220
'ayyil millû'îm	fattened ram	21
'ēyl millu'îm	fattened ram	215
'ēypâh	measure	219
'alyâh	sheep's tail	66, 214
'ănāpâh	heron	150
'ănāqâh	reptile	159
anše-zi-zi (Akk)	horse	89
'aqqô	wild ungulate	186, 218
'ărgāmān	purple	177
'argamannu (Akk)	purple	180
'urwâh	stall	129
'uryâh	stall	129
'orḥat yišmĕ''ēlîm	caravan of Ishmaelites	112
'ărî	lion	198
'aryēh	lion	198
'arnebet	hare	192
'atôn	donkey (f)	97

bĕhēmôt	hippopotamus (?)	195
bôqēr	herder, breeder	46, 82
bikrâh	female camel	130
baqbuq dĕbâš	honey jar	163
bāqār	large cattle	18, 73, 77, 123, 214
ben bāqār	calf	77
ṣemed bāqār	pair of oxen	131
barburîm ʾăbûšîm	fattened birds	21, 154, 164
bāśār	meat	57
bat yaʿănâh	ostrich	150
gĕbînâh	cheese	55, 56
gĕdî	kid	18, 63
gĕdî ʿizzîm	young goat	57, 63
gidrôt ṣoʾn	pens	45
gôzēz	shearer	71
gēz	wool shearing	70
gizzâh	sheared wool	70
gôzāl	chick, young bird	164, 219
gûr	cub	208
gēy ṣĕboʾîm	Valley of Hyenas	205
gĕlômēy tĕkēlet	purple garments	178
gālāl	dung	58
gāmāl	camel	112
dāʾâh	falcon	149
dob	bear	201
dob šakkûl	bereaved bear	201
dĕbôrâh	bee	161
deber	pestilence	68, 76
dĕbāš	honey	52, 161
dĕbāš nĕkoʾt	aromatic honey	163
dāgāh	fish (coll)	19, 167
dayyāgîm	fishermen	168
dûkîpat	hoopoe bird	150
dônag	wax	163
dayyâh	falcon	149

dîšôn	wild ungulate	186, 218
domen	dung	58
zĕʾēb	wolf	203
zibdeh (Ar)	dairy product	54, 55, 83
zĕbûb	fly	161
zemer	wild ungulate	186, 218
zĕroʿâ	forearm, foreleg	228
ḥăzeh	breast	20
ḥăzîr	pig	140
ḥaḥ	hook, nose ring	90
ḥakkâh	fishing implement	168
ḥālāb	milk	52, 83
ḥēleb	animal fat	59
ḥoled	mole	159
ḥemʾâh	dairy product	54, 55
ḥamôr	donkey	90, 127
ḥomet	reptile	158
ḥomeṣ	vinegar	55
ḥēmet	skin container	64
ḥăsîdâh	stork	150
ḥarîṣ hālāb	dairy product	55, 56
ḥērem	fishing net	168
ṭĕrēpâh	devoured animal	20
yônâh	dove	19, 151
yaḥmûr	wild ungulate	186, 218
yam sûp	Sea of Reeds	101
yanšûp	owl	150
yaʿrat haddĕbaš	honeycomb	162
kebeś	sheep	18, 51, 68, 215
kĕbâśîm	sheep (pl)	68
keśeb	sheep	18, 84
kôs	owl	150

miqneh	cattle	51
marḥešet	griddle, pan	20
měrî'	fattened bull	20, 125, 214, 215
merkābâh	chariot	89
mārāq	broth	20, 58
mištaḥ ḥărāmîm	drying place (for nets)	168
meteg	bit, bridle	90, 97
no'd	leather container	64
nēbel	leather container	64
něbēlâh	carcas	20
nôqēd	sheep breeder	49
nāmēr	leopard	198, 201
nopek 'ărgāmān	precious purple dye	178
nēṣ	hawk	150
nešer	griffon vulture	150
sûs	horse	99
sûs wěrokbô	horse and rider	101
sîrôt dûgâh	fishing boats	168
salfih (Ar)	dairy product	55
solet	fine flour	219
sěmādār	grape type	209
semneh (Ar)	dairy product	55
sēpel	bowl	54
'ăbodat 'ădāmâh	farming, land tilling	13
'ăbôt	rope	90
'ēgel	calf	20
'ēgel marbēq	fattened calf	20, 77
'eglâh	calf (f)	124
'ăgālâh	wagon, cart	89
'ēder	herd	51
'ôp	fowl (coll)	19, 151, 164
'ôr	skin, leather	64

ṣayid	hunting, food	206
ṣĕlôḥît	bowl, dish	20
ṣallaḥat	bowl, dish	20
ṣĕlî ʾēš	roast	20, 58
ṣilṣal dāgîm	fishing harpoon	168
ṣmd ḥmrm	pair of asses	95
ṣemer	wool	70
ṣinnâh	boat	181
ṣippôr	bird	164
ṣapîr	kid	62
ṣĕpîr ʿizzîm	young goat	62
qāʾāt	pelican	150
qôrēʾ	partridge	151
rāʾāh	falcon	149
rĕʾēm	wild ox	190
ribqâh	tied ruminants (for milking)	44, 53
rĕḥaṭîm	troughs	47
rôʿeh	herder	48, 49
rāḥāl	ewe	44
rĕḥēlîm	ewes	44
raḥil (Ar)	wandering (for grazing)	44, 130, 216, 218, 218, 228
riḥleh (Ar)	wandering	44, 49
rāḥām	Egyptian vulture	150
rāḥāmâh	Egyptian vulture	150
rekeb	chariot, draft horse	89, 102
rekeb gāmāl	riding camel	116
rekeš	draft horse	102
resen	rein, bridle	90
rĕpātîm	animal sheds	76
rešet	net	168, 181

taḥaš	crocodile (?)	206
tĕḥāšîm	crocodiles (?)	206
tayiš	he-goat	62
tukkiyyîm	exotic birds	155, 205
tĕkēlet	purple	177, 178, 183
takiltu (Akk)	purple	180
tan	jackal	203, 204
tannîm	jackals	204
tannûr	oven	20
tinšemet	barn owl, reptile	150, 159
taršîš	place-name, type of boat	205, 209

Biblical References

Following the order of books in the Hebrew Bible.

Judges, cont.

8:26	178
10:4	97, 127
13:15	40
13:19	63, 221
14	197, 227
14:8–9	162
14:18	124, 196
15:1	57
15:4–5	204
18	60

1 Samuel

1:24	64
1:25	221
2:13–16	221
6	78, 124
6:7	90
6:14	220
8:11	102
9:12–24	20, 221
10:3	221
11:5, 7	123, 131
11:15	221
13:5	102
13:9	221
13:18	205
14:26–27	162
14:32–34	19
15:3	116
15:9	58
15:15	221
16:2	221
16:11	40, 48
16:19	40
16:20	64

1 Samuel, cont.

16:23	49
17:14–15	48
17:18	55
17:34	40, 46
17:34–36	49
17:34–37	196, 197, 202
17:43	135
17:49	49
22:19	40
24:4	45
24:15	135
25:2	71
25:4	71
25:20	97
26:20	151
28:24	77

2 Samuel

1:6	102
1:19	186
3:8	135
6:13	125
8:4	102
9:8	135
10:18	102
13:23–28	71
13:29	110
16:9	135
17:8	202
17:23	97
17:29	54–5
19:27	97, 127
23:20	197
24:22	123
24:25	221

Credits

The illustrations included in Every Living Thing: Daily Use of Animals in Biblical Times *are reproduced with the permission of the following:*

The Bible Lands Museum, Jerusalem. Figures 2.10, 3.5, 3.13.

© Bildarchiv Preussischer Kulturbesitz, Berlin, and the Staatliche Museen, Berlin. Figure 6.2.

© Bildarchiv Preussischer Kulturbesitz, Berlin, and the Ägyptisches Museum, Berlin. Photo by Fremdvölker. Figure 8.1.

© British Museum, London. Figures 2.5, 2.6, 3.12, 4.4, 7.5.

Collections of the Henry Ford Museum and Greenfield Village, Dearborn, Michigan. Figure 2.11.

Zeev Herzog, Institute of Archaeology, Tel Aviv. Figure 8.5.

Israel Antiquities Authority, Jerusalem. Figures 5.5, 7.6.

Ludwig Mayer Jerusalem Ltd. Figures 2.4, 2.7, 2.10, 3.2, 3.7, 3.9, 4.1, 4.6, 5.1, 5.2, 5.3, 5.6, 6.4, 7.1, 7.2, 7.3.

Credits continued

Professor Amihai Mazar, Institute of Archaeology, The Hebrew University, Jerusalem. Figures 2.1, 8.2.

© Musée du Louvre, Antiquitiés Orientales, Paris. Photos Pierre and Maurice Chuzeville. Figures 2.8, 4.3, 7.4.

The Museum Archives, University Museum, University of Pennsylvania, Philadelphia. Figure 3.1.

The Oriental Institute, The University of Chicago. Figures 4.2, 4.5, 4.7.

Taraxacum, Inc., Washington, D. C. Figures 1.1, 1.2.

Professor David Ussishkin, The Institute of Archaeology, Tel Aviv University. Drawing by Judith Dekel. Figure 3.8.

The Walters Art Gallery, Baltimore. Figure 3.11.

Professor Adam Zertal, Department of Archaeology, University of Haifa. Drawings by Judith Dekel. Figures 8.3, 8.4.

The photographs in figures 2.2, 2.3, 2.9, 3.4, 3.6, 3.10 are by the author.